THE
World
Beyond

SCIENCE FICTION AND HORROR FILMS ON KPHO TV5

By Parker Anderson

Cover and interior design: Kubera Book Design, Prescott, AZ

ISBN: 978-0-692-06992-9

DEDICATION

To the memory of my mother, Darla R. Anderson,
who also loved *The World Beyond*

CONTENTS

ACKNOWLEDGEMENTS

The author wishes to expressly thank Logan Blackwell of Phoenix, Arizona, for his kind permission to use his research for this book. It was Logan who spent many lunch hours and Saturday afternoons sifting through the tv listings in old Phoenix newspapers at the Phoenix Public Library for the purpose of compiling a list of every movie that ever aired on *The World Beyond*! He is a researcher after my own heart, and this book would not exist without him.

Unless otherwise noted, all illustrations are courtesy of KPHO-TV.

FOREWORD

By Stu Tracy

AS A KPHO EMPLOYEE during the 70s, 80s, and 90s, I was privileged to play a very minor part in the history of Phoenix television.

Channel 5 always seemed to be on a tight budget, so along with my weather reporting duties on camera, I was also utilized as the station's voice-over announcer.

Among the announcements I would record on a daily or weekly basis were the opens and closes, and the pre-bridges and post-bridges for commercial breaks for all the movie showcases that we ran. And we ran a lot of them!

So, at the time, I thought of this as just part of the job. I had no idea that some young movie fans would remember some of this work more than 40 years later!

For those who are nostalgic about the old days of Phoenix television, enjoy as Parker Anderson takes you on a trip to The Wor-r-r-l-d Be-y-o-n-d.......

The Westward Ho Hotel in downtown Phoenix in the 1950s. KPHO-TV had its broadcast tower on the roof of the facility in those early days. (Author's collection)

KPHO

PHOENIX IS THE CAPITOL OF ARIZONA, a sprawling metropolitan city of over 1 million people, situated in the dry desert of the Salt River Valley in the central part of the state. It is surrounded on all sides by numerous other smaller cities such as Tempe, Mesa, Glendale, and Scottsdale. These are large metropolitan areas as well, and were separated by vacant land in Arizona's early days. Today, however, with the Valley's growth, infrastructure, and labyrinthian freeway system, you can leave one city and enter another without realizing it has happened. The local governments of the cities surrounding Phoenix have been frustrated for years over the loss of their identities, as the entire area is generally considered to be just "Phoenix" by those who do not live there.

KPHO was originally a radio station, owned by a group of Valley businessmen under the name of Phoenix Broadcasting. But as television made its debut in America in the late 1940s, KPHO-TV first signed on the air on December 4, 1949. As with all television stations in the nation that were sprouting up, KPHO broadcast over the free airwaves—things like cable tv and home computers, taken for granted today, were decades in the future. KPHO was the first television station in Arizona.

For its first few years, KPHO carried a variety of programming from the national networks of CBS, NBC, ABC, and the long-defunct DuMont Television Network, as well as some locally produced programs. Like most tv stations of the era, KPHO produced its own programming for children, starting out with the *Lew King Ranger Show*, broadcast in the late afternoons after the kids got home from school. This was soon replaced with the *Gold Dust Charlie Show*, which starred Ken Kennedy (brother of movie director Burt Kennedy) as host, portraying a humorous old prospector.

On his show, Gold Dust Charlie eventually picked up a comic sidekick named Wallace Sneed, portrayed by a young upstart performer named Bill Thompson. When Ken Kennedy left the show in 1954, Thompson took over and a new kiddie show entitled *It's Wallace* was born. He was soon joined by Ladimir "Ladmo" Kwiatkowski and announcer Pat McMahon, and the name was changed to *The Wallace and Ladmo Show*. Wallace and Ladmo are perhaps what KPHO is best remembered for—unlike other kid's show hosts in other states, they became legendary figures in their own lifetimes, and their show (considered rather radical for a children's show) is remembered as a major part of 20th century Arizona history. The program ran uninterrupted on KPHO from 1954–1989, a total of 34 years.

The Meredith Corporation purchased the KPHO television and radio stations in 1952, and own KPHO-TV to this day. Under their management, KPHO went completely independent in 1956—no more borrowings from the national networks. As an independent broadcast station, they now had to produce all of their programming on their own. This would include local news broadcasts, so-called "homemaker" shows for housewives (these were the days before the women's liberation movement), and movies—lots of movies!

Hollywood had initially felt very threatened by the birth of television, but after a few years, the big motion picture studios realized there was money to be made by selling their old movies to television in syndicated packages. Later, this strategy would expand to include reruns of popular tv shows that had gone off the air. For decades, KPHO purchased many of these syndicated packages, and their programming actually rivaled some of today's cable stations.

For film buffs, KPHO became movie heaven in Arizona, with at least one movie shown per day, usually more. On Saturdays, some of the programming was 90% movies for a time. In the days before videos, DVDs, or Netflix, movie lovers would scan that week's issue of *TV*

The cast of KPHO's beloved Wallace and Ladmo Show. From left to right: Ladimir "Ladmo" Kwiatkowski, Pat McMahon (as Gerald), and Bill "Wallace" Thompson. (Courtesy of Steve Hoza at www.wallacewatchers.com)

GUIDE to see if anything they wanted to see was coming up! Older men and women, who grew up in this era (including myself), will tell you there was a certain magic to all of this that today's easy access to movies has not been able to capture—a magic that is unexplainable to anyone who did not live it. You had to have been there in order to understand.

Broadcast stations that ran old movies usually did so with a certain amount of "character" in those days. They would actually give names to certain movie programs, and some of these programs had themes—a certain movie program might only show westerns, for instance. Over the years at KPHO, many movie programs came and went, some lasting for many years, others fading quickly, depending on local ratings. KPHO's movie shows had names like *Action Theatre, Adventure Theatre, Million Dollar Movie, Movietime, TV5 Late Show* (which was later followed by *TV5 Late Late Show!*), and so on.

Such movie programs on local stations introduced them with an appropriate image or graphic, with the show's name on it, while an announcer from the station would welcome viewers and usually spoke the name of the movie and a couple of its stars. Playing in the background would be some music, often cribbed from obscure (and occasionally popular) artists and albums of the era. For instance, KPHO utilized the "Tara Theme" from *Gone With The Wind* for its *Million Dollar Movie*. Both Gershon Kingsley's *Hey Hey* and The Bangles' *Walk Like An Egyptian* were used at different times for KPHO's *Action Theatre*. Part of Peter Thomas' soundtrack for *Chariots Of The Gods* was used for a while by KPHO for *Adventure Theatre*.

Did KPHO and other tv stations get permission and pay expensive royalties to use these samplings? Possibly, but I would tend to doubt it. This was a much different era, and it was a lot easier to fly under the radar in these cases, PLUS in those days, the music and entertainment industries were nowhere near as paranoid about copyright violations as they are today. Today, studios actually hire computer geeks to surf the Internet daily to sniff out possible unauthorized use of their holdings.

But in the 1960s, 70s, and even the early 80s, this worry was not a priority for them. So tv stations were able to use snatches of music to introduce their shows without having to worry much about it.

Movie programs on local stations often utilized "bumpers" at commercial breaks, in which the station announcer would inform the viewer that they would be returning to the movie in a moment. Overall, this kind of personalization by the independent broadcast stations made viewers feel like valued friends, which is something that has been permanently lost in today's easy access to movies and tv shows. Most cable and pay tv outlets (to say nothing of Netflix and other Internet access to films) do not even utilize announcers anymore. They just start the movies.

~

STU TRACY

Stu Tracy was born in Medford, Oregon in 1943. According to an interview he gave me on May 12, 2016, he was working at a local tv station in 1968 when one of his co-workers moved to Phoenix to become manager of KBUZ radio. The man thought the young Stu had announcer possibilities and asked him to come to the Valley and join him there as an announcer.

After about a year of working at KBUZ, Stu Tracy was hired by the more prominent KPHO Radio to do their announcements on December 1, 1969. Soon after, they transferred him to the television station, where he became the voice of KPHO, and he stayed there for over 31 years.

As announcer for KPHO-TV5, Stu's voice was heard all day long. He was required to do station identification at the top of every hour and between programs (a dramatic "KPHO TV5, Phoenix"), as well the movie introductions, promotions for KPHO, sometimes commercials, the announcements for the testing of the Emergency Broadcast System, and anything and everything else that the station needed an announcer

This ad from c. 1971 shows that Stu Tracy had joined KPHO's news staff in some manner already, though he had not yet become the station's weatherman.

for. The voice of Stu Tracy soon became one of the most recognizable in Arizona, even though few television viewers knew his name.

Each night, Stu Tracy would record the next day's announcement

Stu Tracy becomes KPHO's meteorologist.

on reel tape cartridges, which were then played in sequence by the studio's technicians all day and well into the night. For weekends, he would record the movie introductions and all other announcements for both Saturday and Sunday at once. Although given a script, he would, for the commercial break bumpers where he told viewers the film would resume shortly, try to list different stars on them so each bumper/announcement would not sound the same—a striking difference from many tv and radio announcers who just went through the motions.

For well over two decades, the voice of Stu Tracy was familiar to every Arizona resident who regularly watched tv. To movie watchers especially, who tuned in to KPHO regularly, there was a comforting quality to the knowledge that he was always there.

After doing this for quite a few years, KPHO retired Art Brock, the weatherman on its local newscasts, apparently because he was starting to look rather aged on the air. The station proceeded to give the job to Stu Tracy, even though he was not a meteorologist.

Back in those days, surprisingly, most local tv weathermen in America were not meteorologists and largely cribbed the information they passed on from National Weather Service reports. KPHO wanted Stu to have a little more credibility than that, so they had him take a four-year correspondence course in meteorology from the University of Mississippi, which would get him a certificate in the topic, albeit not a degree.

The local television landscape started to change in the late 1980s, as cable tv was becoming King of the Hill. Big cable networks started buying up the exclusive broadcast rights to anything that was worth

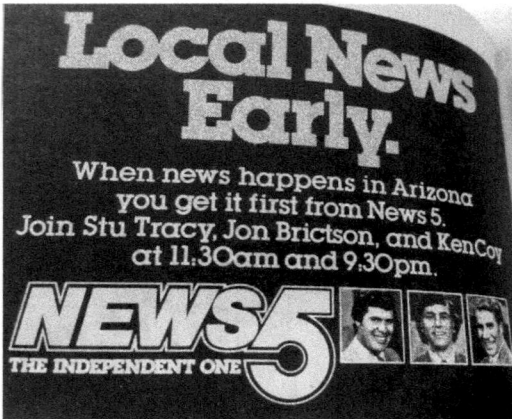

seeing, including old movies and popular tv show reruns. As a result, local stations such as KPHO were forced to start cutting back on this kind of programming. This started to leave Stu Tracy with less to announce.

In 1994, the national network CBS acquired KPHO, and the station became a CBS affiliate (CBS had previously been with KOOL/KTSP-TV in Phoenix). Thus, KPHO ceased to be the independent station it had always been, and no longer had to produce much in the way of local programming, except for its news broadcasts.

Shortly after the CBS acquisition, the new station manager pulled Stu Tracy from what little remained of his announcer duties, apparently out of concern that having an employee do more than one job made them look cheap. Stu continued on as weatherman on the local news broadcasts until 2001, when he retired by mutual agreement.

The legacy of Stu Tracy in Arizona broadcasting is significant, yet underrated, probably because while everyone heard his voice, few knew his name. He was a true original, a man who brought character and enthusiasm to his job as a tv announcer at a time when many similar announcers at stations nationwide just sleepwalked through the job.

Today, Stu Tracy lives in retirement in the Valley, where he pursues his hobby of aviation.

Sadly, much of Stu's announcer output has not survived. In those days, once a recording had been used, it was assumed no one would ever need it again or want it. The tape cartridges were reused and recorded over on. None were kept.

But some have indeed survived.........I was kind of a strange child, and many said so. But I became enamored of movies at an early age, and these were the days I referred to earlier; there were no VCRs, DVDs, or any of the modern gadgets that people take for granted today. Anyway, I developed a very strange hobby around 1975......I took to recording (with an audiocassette recorder) the movie introductions (and sometimes commercial break bumpers) from broadcast stations, and I really enjoyed playing them back! My favorites were, needless to say, from KPHO, where I loved watching *The World Beyond* and other movie shows. I enjoyed this recording so much I continued to do it into my young adult years.

I never told anyone in later years I had this weird hobby when I was growing up, because I figured they would never understand. But recently I became aware of nostalgia in Arizona for *The World Beyond*; nostalgia which led to the book you are now holding. I contacted Logan Blackwell, who has a *World Beyond* Tribute page on Facebook, and I sent him my old cassettes of Stu's World Beyond introductions, not even sure they would play anymore because of their age. He had them digitized and they can now be heard on YouTube! On a small scale, Stu Tracy's broadcast legacy does indeed live on!

For *The World Beyond*, Stu Tracy dramatically intoned "Join us now...." and finished with the words "as we enter into The World Beyoooond!!! (with reverb added). Fans of *The World Beyond* loved his approach, and it is one of the key memories we have. For instance, he might say "Join us now with Gerald Mohr and Les Tremayne in THE ANGRY RED PLANET, as we enter into The World Beyoooond!!!!" It still is effective in the surviving recordings.

～

Horror and sci-fi films had been popular in motion pictures almost since the beginning. When television was born, and such films started being seen by a wide audience at home, it spurred a new revival of motion-picture production of them in the 1950s—mostly low-budget and not highly regarded critically, but very popular. Suddenly, space aliens were everywhere, along with teenage werewolves, and the like. The magazine *Famous Monsters Of Filmland* was born in 1958 which glorified such films and the juvenile audience it was aimed at ate it all up! The editor of *Famous Monsters*, Forrest J. Ackerman, became a legendary figure in his own lifetime because of it.

The nation's independent tv stations, which were now playing syndicated packages of movies daily, created various movie shows specifically designed for horror/science-fiction movies. A number of stations, including KTTV in Los Angeles, utilized the name *Creature Features*. KTLA in Los Angeles ran *Wide Scream Theater* and *Monster Rally* for many years. Oftentimes, stations with these movie shows hired on-screen hosts to introduce the movies, dressed in "creepy" costumes and hamming it up for the kiddies.

KPHO in Phoenix also created movie shows with horror themes in which to broadcast syndicated films from this genre. The station did not utilize live hosts as stations in other states did. Long-time Arizona residents remember TV5's *Chiller* on Saturday evenings that ran for a time, and also *STUDIO 5* which ran for many years on Friday and Saturday late at night. The Saturday night edition of *Studio 5* generally ran mundane, "ordinary" movies, but the Friday night edition of that program, which immediately followed KPHO's *Friday Night At The Movies*, showed horror films 90% of the time, despite the fact that the show's title had no horror inference.

Then......there was *The World Beyond*...............

⌣

KPHO-TV, Channel 5 in Phoenix, debuted its movie program, *The World Beyond*, on September 18, 1964. It ran twice each weekend; first on Friday evenings at 6:30pm, and then again on Saturday afternoons at 1:30pm. Its purpose was to show science-fiction films and "monster movies," and sometimes horror films (in later years, horror films came to dominate more). Most of these movies in the early years were the kind that had been seen on drive-in screens in the 1950s, as well as some older films such as the Universal "monster" classics. These films were pretty innocuous, and the show was a hit with Arizona children, especially those who were growing up on magazines like *Famous Monsters*! Times would invariably change and, in its later years, *The World Beyond* would occasionally show more adult offerings also.

Interestingly, *The World Beyond* was not an original title. The previous year, on October 17, 1963, WTTG Channel 5 in Washington D.C. began a monster/science-fiction movie program with the same name. As WTTG and KPHO were not related, the show title was perhaps a coincidence. But more likely, someone at KPHO had heard of the program in Washington D.C. and decided the title would work for Phoenix too! We will never know for sure, but again, in that era no one really cared about such small "borrowings," especially in this case when the two tv stations were almost a full continent apart.

The first film on WTTG's *World Beyond* appears to have been *The Giant Behemoth* (1959). We are reasonably certain the show ended on August 9, 1966, slightly under three years. By contrast, KPHO's *World Beyond* would last for 24 years! During this nearly quarter of a century, the movies shown were wide and varied, ranging from rock-bottom slop like *THE CREEPING TERROR* (1964) and *CURSE OF THE FACELESS MAN* (1958) to major studio releases like *WESTWORLD* (1973) and Robert Altman's *COUNTDOWN* (1968)! Whenever the station acquired a new movie package, program schedulers had to sort out material that might fit *The World Beyond*; it didn't matter who made it!

⁓

During the years I was actively watching *The World Beyond* and recording the introductions (between 1975 and 1988, now available on YouTube, as noted before), the show went through four theme music changes, though it undoubtedly had earlier ones as well. For the first one after

I started recording, KPHO was using the opening strains from Carlos Santana's *Black Magic Woman*, which was quite a good choice for the show! The following two I cannot identify, although the third is some kind of disco rock with the cries of Godzilla and other Japanese monsters edited in. The fourth and final theme was the opening strain of Vangelis' *Curious Electric.*

Many fans of *The World Beyond* are adamant that KPHO used Pink Floyd's *Time* as the theme for the show for a very prolonged period. This is referenced in numerous Arizona-based blogs that are nostalgic for the show, but there is no evidence of this from the surviving recordings. If it was used, it had to have been before 1975, which seems unlikely since *Time* was issued only in 1973. It is hard to believe that so many different people could have the same false memory—even my younger brother remembered *Time* as the theme for *The World Beyond*—but I have to think it is indeed some sort of mass memory fluctuation. For the record, I have a personal memory, which I cannot prove, that KPHO utilized *Time* on commercials advertising their reruns of *The Outer Limits* tv series. I have no recollection myself of ever hearing it as a *World Beyond* theme. Could so many people really have confused the two in their memory banks? While the odds against that seem great, I have little choice but to believe this is what happened. I am sure many fans will dispute me on this, and I'm sorry for that, as I intend no offense.

On to the movies!

THE MOVIES

ONCE AGAIN, A REMINDER that *The World Beyond* started out airing on Friday evenings at 6:30, and was repeated Saturday afternoons at 1:30.

1964

September 18 and 19—RETURN OF THE FLY (1959) First sequel to the famous THE FLY (1958) finds the son of the scientist from the first film conducting similar experiments with the exact same results. With Vincent Price and Brett Halsey.

September 25 and 26—GIGANTIS, THE FIRE MONSTER (1959) Japanese film was actually the first sequel to the original GODZILLA, but was retitled by the American distributor, possibly due to Copyright issues. It is also known as GODZILLA RAIDS AGAIN. Interestingly in later years, when Godzilla movies became more prominent on *World Beyond*, this one was not among them.

The very first newspaper ad for The World Beyond. *The bottom caption audaciously reads: "The most imaginative, most absorbing science fiction films on a wide variety of themes. Selections from the finest producers here and abroad...truly outstanding entertainment for the whole family."*

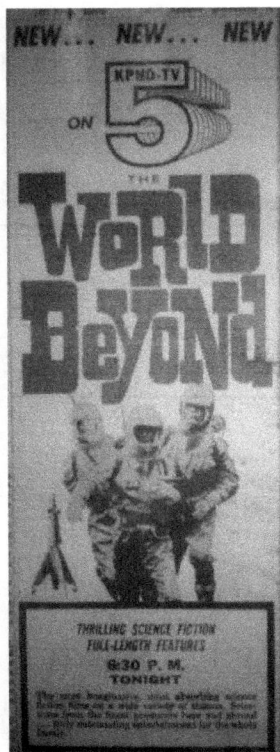

October 2 and 3—FOUR SIDED TRIANGLE (1953) Two scientists in love with the same woman decide to make a clone so they can both have her. With Barbara Payton, James Hayter, Stephen Murray. An early effort from Hammer Films before they hit their stride; directed by Terence Fisher. A film that is not commonly seen today.

October 9 and 10—SATELLITE IN THE SKY (1956) Astronauts rush to disarm a bomb that is threatening to blow up their spaceship. With Kieron Moore, Donald Wolfit, and Lois Maxwell (who would later become a regular in the James Bond series). Another film not widely seen today.

October 16 and 17—X THE UNKNOWN (1956) A giant radioactive blob threatens to destroy mankind in this early outing from Hammer Films, written by Jimmy Sangster. Not bad of its kind. With Dean Jagger, Edward Chapman, Leo McKern, Anthony Newley (before he hit the big time).

October 23 and 24—SPACEWAYS (1953) Another early Hammer film directed by Terence Fisher, in which a scientist is forced to go into space to retrieve a satellite that will prove him innocent of murder. With Howard Duff and Eva Bartok.

October 30 and 31—GORILLA AT LARGE (1954) A vicious gorilla is suspected of murdering people at a carnival sideshow. This was originally released in theatres in 3D. It is

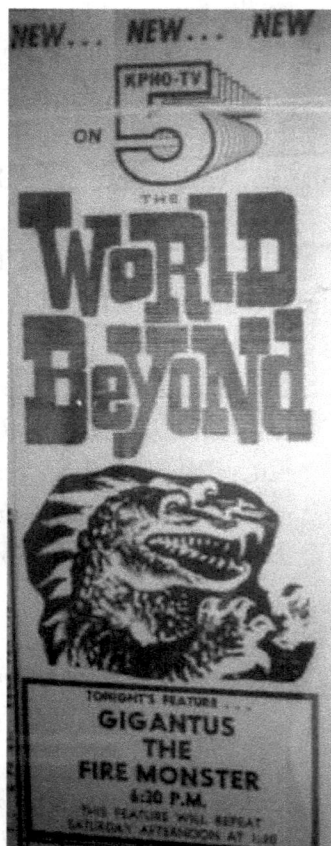

not very good, but some genre fans are fond of it because the cast consists of a number of famous stars before they hit the big time. With Cameron Mitchell, Anne Bancroft, Lee J. Cobb, Raymond Burr, Lee Marvin, Billy Curtis. Unless you want to count RETURN OF THE FLY (which was more sci-fi), this was the first actual "horror" film to be shown on *The World Beyond*.

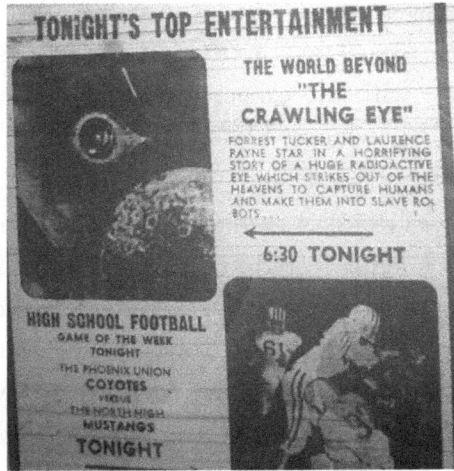

November 6 and 7—THE COSMIC MONSTERS (1958) A mad scientist unleashes giant insects upon the world. With Forrest Tucker, Martin Benson.

November 13 and 14—THE CRAWLING EYE (1958) Space aliens who look like giant eyes with tentacles invade Switzerland. The screenplay was by Jimmy Sangster. With Forrest Tucker, Laurence Payne, Janet Munro.

November 20 and 21— THE ELECTRONIC MONSTER (1957) Detective investigating a murder discovers a clinic where doctors are conducting unscrupulous experiments on their patients. With Rod Cameron and Mary Murphy.

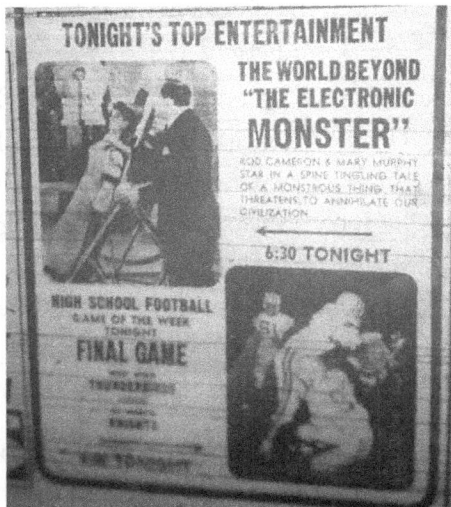

November 27 and 28—SON OF DR. JEKYLL (1951) The title character tries to clear his father's name by recreating his formula, with the expected disastrous results. With Louis Hayward, Jody Lawrence, Alexander Knox.

December 4 and 5—IT CAME FROM BENEATH THE SEA (1955) A giant monster octopus invades San Francisco. Well-remembered by monster fans because of the impressive Ray Harryhausen special effects, which made this film above average. With Kenneth Tobey, Faith Domergue.

December 11 and 12—THE GAMMA PEOPLE (1956) In an unidentified banana republic, two men stumble upon a group of mad scientists who are creating mutant henchmen for the country's ruler. With Paul Douglas, Eva Bartok, Walter Rilla.

December 18 and 19—ONE MILLION B.C. (1940) In prehistoric times, tribes of cavemen war on each other and battle dinosaurs. This ambitious (for its era) film was produced and directed by father and son Hal Roach and Hal Roach Jr. With Victor Mature, Carole Landis, Lon Chaney Jr. A long-standing urban legend that this was partially directed by D.W. Griffith is now believed by film historians to be untrue. As this film is neither horror nor science-fiction, the dinosaurs have to be the only reason this was shown on *The World Beyond*.

December 25 and 26—THE INCREDIBLE PETRIFIED WORLD (1958) Scientists use a diving bell to explore the ocean depths and discover catacombs with a lost civilization. Produced and directed by Jerry Warren, whose low-budget films are regarded as some of the worst movies ever made (this one was slightly above average for him). With John Carradine, Robert Clarke, Phyllis Coates.

1965

January 1 and 2—THE GIANT GILA MONSTER (1959) The title creature goes on a rampage in this ultra-low budget film directed by Ray Kellog and produced by actor Ken Curtis (who played Festus on the Gunsmoke TV series). With Don Sullivan, Shug Fisher.

January 8 and 9—THE UNDEAD (1957) A young woman is transported back to Medieval times where she is accused of being a witch. The first film directed by legendary filmmaker Roger Corman to be shown on *The World Beyond.* The screenplay was co-written by Corman regular Charles B. Griffith. With Pamela Duncan, Allison Hayes, Bruno VeSota, Mel Welles, Billy Barty, Dick Miller. Fans of low-budget horror films could hardly lose with a cast and crew like that!

January 15 and 16—VOODOO WOMAN (1957) Mad scientist in the jungle turns a woman into a monster. What else would he be doing there? Produced by Alex Gordon (who always responded angrily to criticism of any low-budget movies) and directed by Edward L. Cahn. Paul Blaisdell created the monster suit. With Tom Conway, Marla English, Mike Connors.

January 22 and 23—THE SPIDER (1958) Teenagers battle a giant mutant spider. Produced and directed by Bert I. Gordon. With Ed Kemmer, June Kenny. Also known as EARTH VS. THE SPIDER.

January 29 and 30—VIKING WOMEN AND THE SEA SERPENT (1957) The title about says it all in this Roger Corman-directed opus. With Abby Dalton, Susan Cabot, Betsy Jones-Moreland, Jonathan Haze, June Kenny.

February 5 and 6—THE KILLER SHREWS (1959) The title monsters pick off people on a remote Island. Directed by Ray Kellog. With James Best, Ken Curtis (who also produced), Ingrid Goude, Baruch Lumet (Sidney Lumet's father in real life).

February 12 and 13—THE BEAST WITH A MILLION EYES (1955) Space monster terrorizes a family living isolated in the desert. Paul Blaisdell created the monster for the film. With Paul Birch, Chester Conklin (of all people!).

February 19 and 20—THE DAY THE WORLD ENDED (1956) Human survivors vs. mutants (including another one of Paul Blaisdell's monsters) following a nuclear war. Produced and directed by Roger Corman. With Richard Denning, Lori Nelson, Paul Burch, Mike Connors, Jonathan Haze.

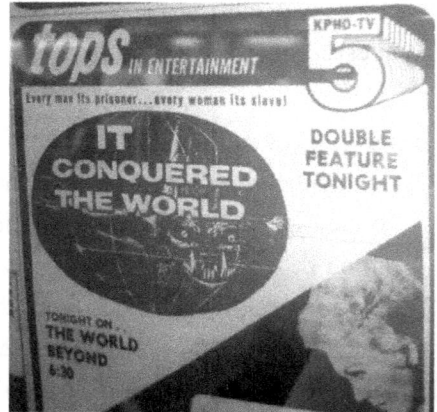

February 26 and 27—THE SHE CREATURE (1956) Mad scientist discovers a beautiful woman is the reincarnation of a prehistoric sea monster, proceeds to transform her into her former self! Yes, you read that correctly. Paul Blaisdell created the monster again. Produced by Alex Gordon and directed by Edward L. Cahn, with a very strange cast including Chester Morris, Marla English, Tom Conway, Ron Randell, Cathy Downs, Frida Inescort, El Brendel, Jack Mulhall.

March 5 and 6—IT CONQUERED THE WORLD (1956) Another space monster (created by costume-maker Paul Blaisdell) attempts to take over the Earth. Produced and directed by Roger Corman, and this film is a favorite among his fans. With Peter Graves, Beverly Garland, Lee Van Cleef, Jonathan Haze, Dick Miller.

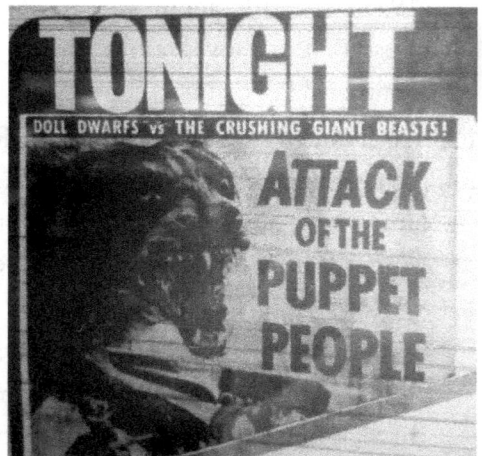

March 12 and 13—ATTACK OF THE PUPPET PEOPLE (1958) A Gepetto-like toy maker kidnaps people, shrinks them, and holds them prisoner. Produced and directed by Bert I. Gordon. With John Agar, John Hoyt.

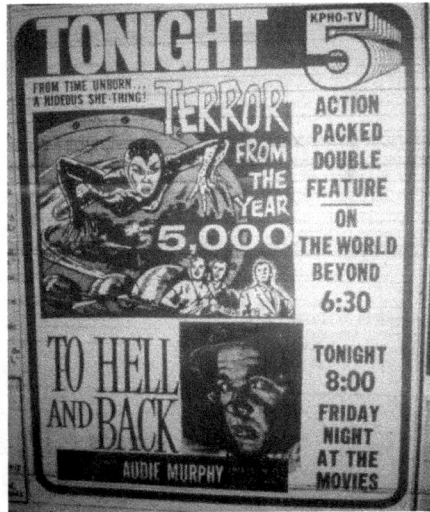

March 19 and 20—TERROR FROM THE YEAR 5000 (1958) Mad scientists fooling around with a time machine bring a sexy woman back from the future, who proceeds to hypnotize and kidnap males to take back to her own time. As silly as it sounds. With Ward Costello, Joyce Holden.

March 26 and 27—WAR OF THE COLOSSAL BEAST (1958) Sequel to THE AMAZING COLOSSAL MAN (which oddly would not air on *The World Beyond* until 1969), with the title character on a rampage again with the Army trying to stop him. Produced and directed by Bert I. Gordon. With Dean Parkin, Sally Fraser.

April 2 and 3—ZOMBIES OF MORA TAU (1957) Treasure hunters try to salvage a sunken vessel guarded by zombies! More silly stuff produced by Sam Katzman and directed by Edward L. Cahn. With Allison Hayes, Morris Ankrum.

April 9 and 10—CARNIVAL OF SOULS (1962) Woman seemingly returns from the dead after drowning, but is unsure of what happened and finds herself pursued by ghostly apparitions. Very low-budget film which has, in more recent years, developed a strong cult following. It was not known at the time, but this would become one of the most highly regarded films to air on *The World Beyond*. With Candace Hilligoss, Sidney Berger.

April 16 and 17—VARAN, THE UNBELIEVABLE (1958) A Japanese monster movie from Toho Studios, the creators of

Godzilla. For American release, added scenes with Myron Healy were shot by the distributor and edited in. Varan did the usual thing, stomping cities and causing mayhem, but he did not catch on the way Godzilla did and there were no sequels.

April 23 and 24—CREATION OF THE HUMANOIDS (1962) Following a nuclear holocaust, the small number of surviving humans have robot servants to do their work, but of course things go horribly wrong! Supposedly this was one of Andy Warhol's favorite films! With Don Megowan, Dudley Manlove.

April 30 and May 1—THE 27th DAY (1957) Sci-fi propaganda film from the Cold War/Red Scare era, as a space alien gives five Earth people capsules which, if opened within 27 days of each other, will destroy the entire planet. In the end, only the Commies are destroyed and the free world lives on! With Gene Barry, Valerie French, George Voskovec.

May 7 and 8—FIRST SPACESHIP ON VENUS (1962) German film, about a manned flight to Venus discovering the remains of a dead alien civilization, was considered a major production in Europe, but it was cut in half by its American distributor and released on the grindhouse/drive-in circuit as just another space opera. Hence its inevitable airing on *The World Beyond*! With Yoko Tani, Oldrich Lukes. Based on a novel by Russian sci-fi author Stanislaw Lem, who disavowed the film.

May 14 and 15—DOCTOR X (1932) Authorities try to apprehend a strangler known as the Full Moon Killer before he strikes again! Early sound film was the oldest movie to air on *World Beyond* up to that time, and is a favorite of fans of 1930s horror movies, though it has dated very badly. Directed by Michael Curtiz, who would later go on to direct CASABLANCA. With Lionel Atwill, Fay Wray, Lee Tracy, Preston Foster, Mae Busch.

May 21 and 22—THE MAGNETIC MONSTER (1953) A radioactive isotope grows and threatens the world! Produced by

Ivan Tors and directed by Curt Siodmak, and written by them both. With King Donovan and Richard Carlson. Much of the action is reportedly stock footage from a 1934 German film called GOLD which was never released in America.

May 28 and 29—THE VAMPIRE (1957) Small town doctor ingests a drug made from the blood of a vampire, turns into a monster, of course. Rereleased at some point under the alternate title of MARK OF THE VAMPIRE (no relation to the famous 1935 film of the same name), and would later be rerun on *World Beyond* in 1978 under this name as well. With John Beal, Colleen Gray, Dabbs Greer.

June 4 and 5—12 TO THE MOON (1960) Expedition to the moon discovers a race of space aliens bent on destroying Earth. Routine, needless to say. With Ken Clark, Tom Conway, former silent film star Francis X. Bushman.

June 11 and 12—THE MAN FROM PLANET X (1951) Space alien lands in the Scottish Highlands. At first he is peaceful, but nefarious humans turn him against us! Directed by Edgar G. Ulmer. With Robert Clarke, Margaret Field, William Schallert.

June 18 and 19—FACE OF MARBLE (1946) Mad scientist experiments with trying to bring the dead back to life! This was released by Monogram Studios, a poverty-row outfit that to this day has the reputation of having made some of the worst horror films ever. With John Carradine, Claudia Drake, Robert Shayne, Willie Best.

June 25 and 26—THE LOST MISSILE (1958) An alien missile begins orbiting the Earth, causing mayhem and destruction. With Ellen Parker, Robert Loggia (who went on to better things in his long career).

July 2 and 3—THE HUMAN MONSTER (1939) A doctor at a home for the blind murders patients and collects on their insurance policies. A small-budget British film, based on an Edgar Wallace novel, was titled DARK EYES OF LONDON in England. More of a murder mystery than a horror film, but the American release title and Bela Lugosi as the star cemented its reputation in the horror genre. Interestingly, this was the first Lugosi film to be shown on *World Beyond*.

July 9 and 10—THE NEANDERTHAL MAN (1953) Another mad scientist inadvertently turns himself into a murderous caveman. Directed by E.A. Dupont, who had once been a highly- regarded director of silent films in Germany. With Robert Shayne, Doris Merrick, Richard Crane.

July 16 and 17—THE INCREDIBLE PETRIFIED WORLD (1958) The first *World Beyond* rerun, from December 25 and 26, 1964.

July 23 and 24—ROCKETSHIP X-M (1950) Astronauts land on Mars, battle race of mutants created by radiation. With Lloyd Bridges, Hugh O'Brian, Noah Beery Jr., Morris Ankrum. Very routine stuff, though many sci-fi fans have a high regard for it because it was one of the first movies to utilize this much-overused premise.

July 30 and 31—THE ELECTRONIC MONSTER (1960) Rerun.

August 6 and 7—CREATURE WITH THE ATOM BRAIN (1955) A mad scientist creates an army of robotic men who can't be killed, which are then used by gangsters for their nefarious ends. Written by Curt Siodmak and directed by Edward L. Cahn. With Richard Denning, Angela Stevens.

August 13 and 14—GIGANTIS, THE FIRE MONSTER (1959) Rerun.

August 20 and 21—ATOMIC RULERS OF THE WORLD (1964) One of a series of Japanese movies aimed at children, about a caped superhero named Starman battling various evil aliens. These were sold straight to television in America. Although 1964 is always given as the release date, these films are believed to have been originally shot in the 1950s.

August 27 and 28—THE GAMMA PEOPLE (1956) Rerun.

September 3 and 4—THE LOST CONTINENT (1951) Expedition seeking a crashed rocketship discovers a prehistroric land filled with dinosaurs. With Cesar Romero, John Hoyt, Hugh Beaumont, Whit Bissell. Acquanetta appears in the film as well; she retired not long after this film was made and, ironically, married a Phoenix car dealer and appeared in his tv commercials in the 1960s and 70s!

September 10 and 11—THE INVISIBLE MAN (1933) This was the first of Universal Studio's "classic" horror movies to be shown on *The World Beyond*. It remains one of the highest-regarded horror movies ever made, and turned Claude Rains into a star. Also with Gloria Stuart, Henry Travers, Dudley Digges, Una O'Connor, E.E. Clive, Dwight Frye, and an unbilled bit by John Carradine before he hit the big time.

September 17 and 18—THE DAY THE WORLD ENDED (1956) Rerun.

September 24 and 25—SPACEWAYS (1953) Rerun.

October 1 and 2—DONOVAN'S BRAIN (1953) Scientist's personality is taken over by the sinister brain he is keeping alive in his laboratory. Based on a Curt Siodmak novel that has been filmed many times. With Lew Ayres, Gene Evans, Nancy Davis (who, I'm sure everyone knows, married Ronald Reagan and later became First Lady of the United States!).

October 8 and 9—RED PLANET MARS (1952) Scientists start picking up messages from Mars, which turn out to be from God, warning us of the threat of Communism! A product of the "Red scare" McCarthy era, based on a 1930s stage play by John L. Balderston. With Peter Graves, Andrea King, Marvin Miller, Morris Ankrum, Bayard Veiler.

October 15 and 16—THE INVISIBLE MAN'S REVENGE (1944) Second sequel to the original film, with Jon Hall as the title character this time. Also with John Carradine, Evelyn Ankers, Gale Sondergaard.

October 22 and 23—THE CREATURE FROM THE BLACK LAGOON (1954) The famous original film, with the title beast attacking people in an Amazon jungle. With Richard Carlson, Julie Adams, Richard Denning, Whit Bissell.

October 29 and 30—INVASION OF THE NEPTUNE MEN (1964) Another Japanese film with a flying superhero battling aliens, sold straight to television in America.

November 5 and 6—THE MONSTER THAT CHALLENGED THE WORLD (1957) Monsters that look like giant caterpillars wreak the usual havoc. With Tim Holt, Hans Conried.

November 12 and 13—THE CREATURE WALKS AMONG US (1956) Second and final sequel to CREATURE FROM THE BLACK LAGOON has scientists capturing the "Gill Man" and trying to humanize him. Good luck with that! With Jeff Morrow, Rex Reason.

November 19 and 20—ABBOTT AND COSTELLO GO TO MARS (1953) Bud and Lou take off in a rocket and crash land on Venus (not Mars, despite title), which is inhabited by (who else?) voluptuous women! First actual comedy to be shown on *World Beyond*—although many of the films that had aired thus far had plenty of unintentional laughs!

November 26 and 27—IT CAME FROM OUTER SPACE (1953) Space aliens who have crashed in the desert take over the identities of townspeople in order to make repairs to their ship. With Richard Carlson, Barbara Rush, Russell Johnson. Some fans of 1950s sci-fi films think quite highly of this one (I'm not among them).

December 3 and 4—THE BLACK CAT (1934) After a car accident, young couple becomes trapped in the home of a Satanic priest. Considered one of the finest horror films of the 1930s, the film was quite grisly for its day, and was by far the most "adult" film to be shown on *World Beyond* up to this point. It may have been scheduled simply because it stars Boris Karloff and Bela Lugosi (this was the first Karloff film to be shown on *World Beyond*). Also with David Manners, Lucille Lund, unbilled bit part by John Carradine.

December 10 and 11—ATTACK FROM SPACE (1964) More adventures of the caped Japanese superhero Starman, as he battles more aliens.

December 17 and 18—BATTLE IN OUTER SPACE (1960) Japanese film with Earthmen battling space aliens, with all of them flying around in saucers, shooting at each other.

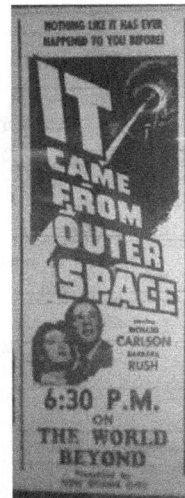

December 24 and 25—*The World Beyond* was apparently pre-empted by a Mahalia Jackson Christmas special and other KPHO Christmas programming in both time slots.

December 31—GODZILLA, KING OF THE MONSTERS (1956) The original Godzilla film, albeit the American release version with added footage of Raymond Burr.

1966

January 1—GODZILLA, KING OF THE MONSTERS (1956) (see above).

January 7 and 8—THE GIANT GILA MONSTER (1959) Rerun.

January 14 and 15—THE BEAST WITH A MILLION EYES (1955) Rerun.

January 21 and 22—THE EVIL BRAIN FROM OUTER SPACE (1964) The Japanese superhero Starman returns again for more silly, toylike adventures.

January 28 and 29—THE H-MAN (1959) An atomic bomb blast turns men into radioactive slime. Unusually dull Japanese-made sci-fi.

February 4 and 5—IT CAME FROM OUTER SPACE (1953) Rerun.

February 11 and 12—GOG (1954) The Communists take over a pair of American-made robots and cause them to kill and go on a rampage. More sci-fi fun from the "Red Scare" era, produced by Ivan Tors and directed by Herbert L. Strock. With Richard Egan, Herbert Marshall, William Schallert.

February 18 and 19—RETURN OF THE FLY (1959) Rerun.

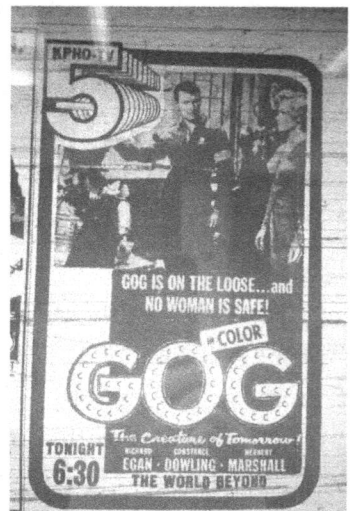

KPHO 5

GOG IS ON THE LOOSE...and NO WOMAN IS SAFE!

IN COLOR

GOG

The Creatures of Tomorrow!

TONIGHT
6:30

RICHARD   CONSTANCE   HERBERT
EGAN · DOWLING · MARSHALL
THE WORLD BEYOND

February 25 and 26—20 MILLION MILES TO EARTH (1957) Space flight to Venus returns to Earth with a small monster that grows…and grows… and escapes, causing mayhem. The script is routine, but the Ray Harryhausen-created monster remains a lot of fun to watch, and has kept this film a favorite among fans of old sci-fi movies. The creature has no name in the film, but his fans have dubbed him "the Ymir." With William Hopper, Joan Taylor.

March 4 and 5—IT CAME FROM BENEATH THE SEA (1955) Rerun.

March 11 and 12—FOUR SIDED TRIANGLE (1953) Rerun.

March 18 and 19—PRINCE OF SPACE (1962) Another Japanese film with a caped superhero battling aliens. Although not part of the Starman series, this film is very similar indeed, aimed at young children.

March 25 and 26—THE EVIL BRAIN FROM OUTER SPACE (1964) Rerun, unusually soon after its January 21 and 22 airing on *The World Beyond*.

April 1 and 2—NIGHT MONSTER (1942) A group of people gathered at an old mansion are picked off one by one. Produced by Universal on a low-budget, this is actually not bad for a routine "old dark house" murder mystery. With Bela Lugosi, Lionel Atwill, Irene Hervey, Don Porter, Leif Erickson, Ralph Morgan.

April 8 and 9—THE FINAL WAR (1960) Japanese-made film depicting the political events leading up to the end of the world in a nuclear holocaust. Not really a science-fiction film at all, but it sounds like one, which is undoubtedly why it was scheduled on *The World Beyond*. The film is difficult to find today.

April 15 and 16—VARAN THE UNBELIEVABLE (1962) Rerun.

April 22 and 23—THE LOST CONTINENT (1951) Rerun.

April 29 and 30—SPACEWAYS (1953) Rerun.

May 6 and 7—GIGANTIS, THE FIRE MONSTER (1959) Rerun.

May 13 and 14—FIRST SPACESHIP ON VENUS (1962) Rerun.

May 20 and 21—THE UNDEAD (1957) Rerun.

May 27 and 28—ATOMIC RULERS OF THE WORLD (1964) Rerun.

June 3 and 4—INVASION OF THE NEPTUNE MEN (1964) Rerun.

June 10 and 11—INVADERS FROM SPACE (1964) Still more adventures of the Japanese superhero Starman, who must have been a popular character, at least in Japan!

June 17 and 18—THE ELECTRONIC MONSTER (1960) Rerun.

June 24 and 25—THE INVISIBLE CREATURE (1960) A ghost saves a woman from her husband who is trying to kill her. With Tony Wright, Patricia Dainton, Sandra Dorne. Film is not easily found today.

July 1 and 2—CREATURE WITH THE ATOM BRAIN (1955) Rerun.

July 8 and 9—THE H-MAN (1959) Rerun.

July 15 and 16—BATTLE IN OUTER SPACE (1960) Rerun.

July 22 and 23—THE LOST BATTALION (1962) During World War II, Philippine and American troops try to outrun Japanese invaders. A routine war movie, made in the Philippines, scheduled on *The World Beyond* for no discernable reason. It can only be assumed that a KPHO programmer mistook it for a horror or sci-fi film in a rushed moment. Directed by Eddie Romero, who made many low-budget films in the Philippines. With Leopold Salcedo, Diane Jergens. Film is scarce and difficult to locate today. An alternate title was ESCAPE TO PARADISE.

July 29 and 30—NIGHT TIDE (1961) Sailor falls in love with a carnival girl, ignoring warnings that there may be something supernatural about her. This film has developed a small cult following over the years. Directed by Curtis Harrington. With Dennis Hopper, Linda Lawson, Luana Anders, Bruno VeSota.

August 5 and 6—X-THE MAN WITH THE X-RAY EYES (1963) Scientist develops a serum that gives him X-ray vision. Directed by Roger Corman. With Ray Milland, Don Rickles (!), Harold J. Stone, John Hoyt.

August 12 and 13—THE DAY THE EARTH FROZE (1959) Russian-made fantasy set in Medieval times, in which an evil witch steals the sun and holds it for ransom. Oddball stuff. With Nina Anderson, Jon Powers (names undoubtedly Anglicized for American television release), narrated by Marvin Miller.

August 19 and 20—THE ANGRY RED PLANET (1959) Astronauts on a spaceship battle Martian monsters. Directed by Ib Melchior. With Gerald Mohr, Les Tremayne, Jack Kruschen. This was the first showing of what would become one of the most rerun films on *The World Beyond*.

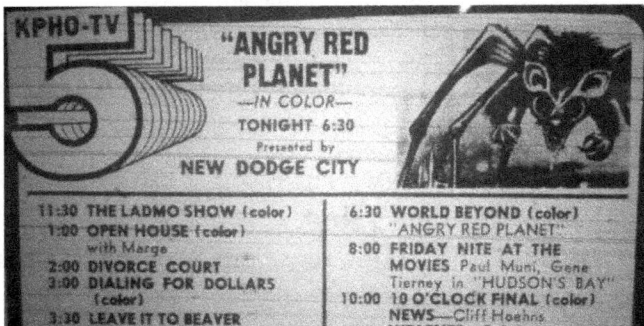

August 26 and 27—BATTLE BEYOND THE SUN (1962) Rival nations try to beat each other to Mars. Originally a Russian-made film, this was re-edited (with some quickly shot new footage) for American release by Roger Corman and Francis Ford Coppola (before he hit the big time). With Edd Perry, Arla Powell.

September 2 and 3—SLAUGHTER OF THE VAMPIRES (1962) Italian film with vampires running amok in an old castle. With Walter Brandi.

September 9 and 10—TERROR FROM THE YEAR 5000 (1958) Rerun.

September 16 and 17—TEENAGERS FROM OUTER SPACE (1959) A teenage space alien falls in love with an Earth girl, and foils the plans of his elders to invade! Not officially a comedy, but there are plenty of unintentional laughs, needless to say! With David Love, Dawn Anderson.

September 23 and 24—ATTACK FROM SPACE (1964) Rerun.

September 30 and October 1—THE ATOMIC MAN (1956) Scientist is exposed to radiation, causing his mind to move 7 seconds into the future, allowing him to answer questions before they are asked and things like that! Naturally, spies try to kidnap him. With Gene Nelson, Faith Domergue.

October 7 and 8—INVASION OF THE STAR CREATURES (1961) Comedy about two bumbling sailors who foil a plot by sexy alien women to take over the Earth! Directed by Bruno VeSota and written by Jonathan Haze. This was apparently a failed attempt to make a comedy team out of stars Bob Ball and Frankie Ray.

October 14 and 15—RIDERS TO THE STARS (1954) Astronauts blast into space to track down potentially dangerous meteors. Produced by Ivan Tors with a screenplay by Curt Siodmak. With William Lundigan, Richard Carlson (who also directed), Herbert Marshall, Martha Hyer, Dawn Addams, King Donovan.

October 21 and 22—THE UNEARTHLY STRANGER (1963) Scientist marries a woman who turns out to be a space alien. Soon, more alien women arrive to take over the Earth. With John Neville.

October 28 and 29—EARTH VS. THE FLYING SAUCERS (1956) Just what the title promises, boosted by terrific Ray Harryhausen-created special effects. With Hugh Marlowe, Joan

Taylor, Morris Ankrum, Paul Frees.

November 4 and 5—20 MILLION MILES TO EARTH (1957) Rerun.

November 11 and 12—ASSIGNMENT: OUTER SPACE (1962) Italian space opera about astronauts who blast off to stop a malfunctioning spaceship from crashing into the Earth. With Archie Savage.

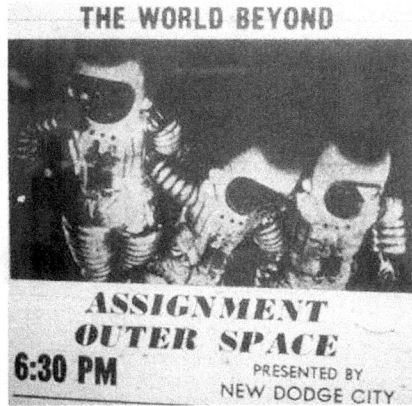

November 18 and 19—BEYOND THE TIME BARRIER (1959) Pilot breaks the time barrier and lands in the future, where he finds the post-nuclear war remains of our civilization. Directed by Edgar G. Ulmer. With Robert Clarke, Darlene Tompkins.

November 25 and 26—CALTIKI, THE IMMORTAL MONSTER (1960) An Italian movie set in Mexico, as an expedition battles a glob-like monster near some old Mayan ruins. With John Merivale, Didi Sullivan.

December 2 and 3—CREATION OF THE HUMANOIDS (1962) Rerun.

December 9 and 10—PRINCE OF SPACE (1962) Rerun.

December 16 and 17—THE DAY THE SKY EXPLODED (1958) Scientists rush to avert a storm of meteors that could destroy the Earth. With Paul Hubschmid, Madeleine Fischer.

December 23 and 24—ONE MILLION B.C. (1940) Rerun.

December 30 and 31—THE JUNGLE (1952) An expedition in the jungles of India discovers a herd of wooly mammoths! With Rod Cameron, Cesar Romero, Marie Windsor, Sulchana.

1967

January 6 and 7—THE BRAIN EATERS (1958) Spongy parasites crawl out of the Earth and attack people, turning them into mindless zombies. Directed by Bruno VeSota. With Ed Nelson, Jody Fair, Leonard Nimoy (long before he hit the big time).

At this point, KPHO-TV cancelled *The World Beyond* and replaced it with *The Gunslingers*, a movie program that aired only Westerns. Why they did this is unknown, but perhaps the program managers at the station thought that Westerns might draw a bigger audience than grade-B (or lower) sci-fi films.

This continued for the remainder of 1967 and into 1968, until February 3, 1968 (a Saturday), when *The World Beyond* returned to the KPHO line-up as mysteriously as it had disappeared, but with a new time slot at 2:00pm and down to a single showing.

1968

February 3—CREATION OF THE HUMANOIDS (1962) Rerun.

February 10—ASSIGNMENT: OUTER SPACE (1962) Rerun.

February 17—WAR OF THE SATELLITES (1958) Space aliens take over a scientist's body to sabotage the space program. Directed by Roger Corman. With Susan Cabot, Dick Miller, Robert Shayne, Bruno VeSota.

February 24—THE ATOMIC MAN (1956) Rerun.

March 2—QUEEN OF OUTER SPACE (1958) Astronauts land on a planet ruled by sexy female aliens (again!). With Zsa Zsa Gabor, Eric Fleming.

March 9 and March 16—*The World Beyond* was apparently pre-empted by a PGA golf tournament.

March 23—12 TO THE MOON (1960) Rerun.

March 30—THE ELECTRONIC MONSTER (1960) Rerun.

April 6—*The World Beyond* was pre-empted by a PGA golf tournament.

April 13—VARAN, THE UNBELIEVABLE (1962) Rerun.

April 20—THE FINAL WAR (1960) Rerun.

April 27—INVASION OF THE NEPTUNE MEN (1964) Rerun.

May 4—THE WOMAN EATER (1959) Mad scientist feeds women to his bloodthirsty tree! With George Coulouris, Robert MacKenzie.

May 11—Apparently *The World Beyond* was pre-empted by a PGA golf tournament again.

May 18—THE 27th DAY (1957) Rerun.

May 25—THE H-MAN (1959) Rerun.

June 1—*The World Beyond* was pre-empted by another PGA golf tournament.

June 8—ATTACK FROM SPACE (1964) Rerun.

June 15—PLANETS AGAINST US (1961) Italian-French co-production about female humanoids with hypnotic eyes who come to Earth to destroy us. Regarded as extremely dull. With Jany Clair, Michele Lemoine.

June 22—THE UNDEAD (1957) Rerun.

June 29—THE ASTOUNDING SHE-MONSTER (1957) A sexy female space alien (another one!) tangles with kidnappers and their victims in the woods. With Robert Clarke, Shirley Kilpatrick.

July 6—BATTLE BEYOND THE SUN (1963) Rerun.

July 13—THE DAY THE EARTH FROZE (1959) Rerun.

July 20—INVASION OF THE STAR CREATURES (1961) Rerun.

July 27—THE INVISIBLE CREATURE (1961) Rerun.

August 3—*The World Beyond* is pre-empted by a PGA golf tournament.

August 10—WAR OF THE COLOSSAL BEAST (1958) Rerun.

August 17—*The World Beyond* is pre-empted by a PGA golf tournament once again.

August 24—THE BRAIN EATERS (1958) Rerun.

August 31—THE DAY THE WORLD ENDED (1956) Rerun.

September 7—THE HYPNOTIC EYE (1960) An evil hypnotist mesmerizes beautiful women and has them go home and disfigure their faces. Considered pretty grisly in its day. With Jacques Bergerac, Allison Hayes, Merry Anders.

September 14—*The World Beyond* was apparently pre-empted by a Kemper golf tournament.

At this point, *The World Beyond* underwent another time slot change, with KPHO moving it to 1:30pm, a half hour earlier.

September 21—NIGHT OF THE BLOOD BEAST (1958) Astronaut who returns to Earth is found to have alien embryos implanted in him (making him an expectant mother!), while the alien itself goes around slaughtering people. Very low-budget and poorly made, but the offbeat plot has given this film a small following in later years. With Ed Nelson, Michael Emmet.

September 28—THE HAND (1960) A London police detective investigates a series of murders in which the victims' hands are cut off. With Derek Bond, Ray Cooney.

October 5—THE ATOMIC SUBMARINE (1959) A submarine crew investigating the mystery of some missing ships discovers they are being destroyed by an underwater flying saucer piloted by an alien who looks like a giant furry eyeball! The bizarre monster and a very strange cast has made this one a favorite among fans of the era's low-budget science-fiction films. Produced by Alex Gordon.

With Arthur Franz, Dick Foran, Brett Halsey, Joi Lansing, Tom Conway, Bob Steele, Victor Varconi.

At this point, while the 1:30pm Saturday time slot stays intact, KPHO starts repeating the movies at midnight also, apparently without *The World Beyond* name for that showing. This did not last long; maybe a year at the most.

October 12—BEYOND THE TIME BARRIER (1960) Rerun.

October 19—NOT OF THIS EARTH (1957) Space alien arrives on Earth to get blood for his dying planet, tricks a nurse into helping him. Directed by Roger Corman. With Paul Birch, Beverly Garland, Jonathan Haze, Dick Miller.

October 26—CALTIKI, THE IMMORTAL MONSTER (1960) Rerun.

November 2—THE MAN FROM PLANET X (1951) Rerun.

November 9—TERROR FROM THE YEAR 5000 (1956) Rerun.

November 16—DONOVAN'S BRAIN (1953) Rerun.

November 23—THE GIANT BEHEMOTH (1959) A radioactive dinosaur comes out of the sea and destroys cities. Pretty slow going until the monster appears, then Willis O'Brien's special effects are a lot of fun. With Gene Evans, Andre Morrell, Jack MacGowran.

November 30—CARNIVAL OF SOULS (1962) Rerun.

December 7—THE UNEARTHLY STRANGER (1963) Rerun.

December 14—RETURN OF THE FLY (1959) Rerun.

December 21—THE MANSTER (1959) Man receives an injection from a mad scientist and grows an extra head. Don't you hate it when that happens? This was supposedly the first two-headed monster movie. With Larry Stanford.

December 28—WORLD WITHOUT END (1956) Astronauts land on Earth in the 26th century and find a cowardly race of old men living underground because they don't want to fight the mutants that now rule the planet. Absurd in all departments. With Hugh Marlowe, Nancy Gates, Rod Taylor (before he hit the big time).

1969

The new year brought a new time slot change for a few weeks, with *The World Beyond* moving to 11:30am on Saturdays, with its repeat showing at midnight still intact. This was apparently due to College Basketball games airing at 1:30pm.

January 4—THE BLACK SCORPION (1957) Giant scorpions invade the Southwest, courtesy of special effects by Willis O'Brien (who had done KING KONG in better days). With Richard Denning and Mara Corday.

January 11—FIRE MONSTERS AGAINST THE SON OF HERCULES (1962) The title hero battles cavemen and some very fake-looking monsters. This was one of a long string of Italian "muscleman" movies that were apparently popular in their home country. With Reg Lewis and Margaret Lee.

January 18—THE PHANTOM PLANET (1961) A film considered to be so bad it's almost fun to watch! Astronaut crashes on an asteroid which is inhabited by people who are 6 inches tall. They shrink him to their size, where he falls for a mute girl, and helps them fight off a warring tribe! With Dean Fredericks, Coleen Gray, Anthony Dexter, Dolores Faith, and Francis X. Bushman, who, in his youth, had been a big star of silent movies, but was one of those who did not survive the transition to sound. (For unknown reasons, the midnight repeat of World Beyond was moved to 1:30a.m. for this one week).

January 25—VENUS AGAINST THE SON OF HERCULES (1963) Another Italian "muscleman" film with the brawny hero taking on a warring tribe (is there any other kind of tribe in the movies?). With Roger Browne and Jackie Lane.

At this point, KPHO moved the Saturday morning airing of *The World Beyond* to 9:30a.m. for several weeks. The midnight repeat stayed the same.

February 1—THE MAGNETIC MONSTER (1953) Rerun.

February 8—THEM! (1954) The military takes on an army of giant ants in the desert! Unlike similar films, this one is remembered quite fondly by sci-fi buffs and is indeed one of the better giant insect movies from the era. With James Whitmore, Edmund Gwenn, James Arness, Onslow Stevens, bit part by Leonard Nimoy when he was just starting out.

February 15—THE FLY (1958) After airing the sequel more than once, *The World Beyond* finally ran the original film. Near legendary sci-fi film with a scientist whose experiments cause him to replace his head with that of a fly! Most of the film's fans have always considered this to be an unintentional laugh riot; by contrast, I have always found it to be rather grisly for some reason. Screenplay by James Clavell. With David Hedison (billed as Al Hedison), Patricia Owens, Vincent Price, Herbert Marshall.

February 22—THE LOST WORLD (1960) Very juvenile version of Sir Arthur Conan Doyle's much-filmed opus about a group of explorers who find a lost land inhabited by dinosaurs. Produced and directed by Irwin Allen, who would become famous for so-called "disaster" movies in the 1970s. With Michael Rennie, Jill St. John, Claude Rains, David Hedison, Fenando Lamas.

March 1—THE THING (1951) Small group of scientists at a scientific research station in the Arctic find themselves menaced by a monster who resembles a giant vegetable (played by James Arness!). Also with Kenneth Tobey, Dewey Martin. Regarded as an all-time classic of its era by most fans; I have never understood this, as I have found it to be not much better than most 1950s sci-fi flicks. It has been remade twice (so far).

At this point there was another time slot change, as KPHO moved *The World Beyond* back to 1:30pm on Saturdays and kept it there for awhile.

March 8—THE ABOMINABLE SNOWMAN OF THE HIMALAYAS (1957) An expedition finds the title monster in this British film with good production values, produced by the fabled Hammer Films. This was the first film with horror movie legend Peter Cushing to air on *The World Beyond*. Also with Forrest Tucker.

March 15—DAGORA, THE SPACE MONSTER (1964) Giant alien monsters attack the Earth in this Japanese movie, produced by Toho. Unlike other Japanese monster movies, these beasts did not catch on and there were no sequels.

March 22—THE MONSTER THAT CHALLENGED THE WORLD (1957) Rerun.

March 29—FRANKENSTEIN 1970 (1958) Very low-budget opus in which Baron Frankenstein attempts to create a new monster while allowing a TV crew to film in his castle. With Boris Karloff, Tom Duggan, Mike Lane. Karloff's career was rapidly sliding downhill at this point, as happens all too often with aging actors.

April 5—*The World Beyond* was preempted by a PGA Golf Tournament.

TONIGHT

5 WILL GET YOU

12:00 EAST-WEST ALL-STAR BASKETBALL GAME (C)
1:30 WORLD BEYOND "FRANKENSTEIN — 1970"
3:30 ROLLER DERBY
4:00 HORSE RACING Live from Turf Paradise.
4:30 MODERN GOLF (C)
5:00 OUTDOORSMAN (C)
5:30 McHALES NAVY
6:00 WESTERN THEATRE (C) Jeff Chandler, Rory Calhoun in "THE SPOILERS." The Yukon wilderness of gold and greed.
8:00 LAREDO (C)
9:00 RAWHIDE
10:00 MOVIETIME (C) Jeff Chandler, Jack Palance in "SIGN OF THE PAGAN." Biblical adventure drama.
12:00 LATE MOVIETIME "FRANKENSTEIN — 1970"

KPHO-TV

A sample of KPHO's broadcast schedule from the era.

April 12—WAR OF THE SATELLITES (1958) Rerun.

April 19—BEAST FROM 20,000 FATHOMS (1953) Dinosaur revived by atomic blast invades New York City with the help of terrific Ray Harryhausen special effects. With Paul Christian, Paula Raymond, Cecil Kellaway, Lee Van Cleef, King Donovan, Kenneth Tobey, Ross Elliott.

April 26—THE AMAZING TRANSPARENT MAN (1959) Mad scientist turns a criminal invisible, but the miscreant has his own ideas on what to do with his new powers. Directed by Edgar G. Ulmer. With Douglas Kennedy, Marguerite Chapman.

May 3—*The World Beyond* was pre-empted by a PGA Golf Tournament.

May 10—SPACE MONSTER (1964) Can a title get any more forgettable than this? Very low-budget opus with still more astronauts meeting still more space creatures. This was written and directed by Leonard Katzman, who would later go on to become a successful TV producer (he created the famous DALLAS series, among other things). Film is very difficult to find today, making one wonder if Katzman bought up all the prints after he hit the big time. With Russ Bender, Francine York, James B. Brown.

May 17—THE GIANT OF METROPOLIS (1962) Another Italian "muscleman" film, with the hero finding a lost city inhabited by the usual "bad guys." With Gordon Mitchell.

May 24—MR. SARDONICUS (1961) An evil Baron forces a doctor to operate on his disfigured face. Produced and directed by William Castle, and a favorite among fans of his "gimmick" films. With Guy Rolfe, Oscar Homolka. (*The World Beyond* aired at 1:00pm this week, following an early PGA Golf Tournament. If the game ran overtime, it is possible that the movie began "in progress," a common practice on broadcast tv with sporting events).

May 31—RIDERS TO THE STARS (1954) Rerun.

June 7—CULT OF THE COBRA (1955) Native cult priestess with supernatural powers seeks revenge on the American soldiers who photographed one of her rituals. With Faith Domergue, Richard Long, Marshall Thompson, David Janssen. (*The World Beyond* aired at 1:00pm this week, probably due to another sporting event).

June 14—INVASION (1964) Another nominee for most forgettable and meaningless title for a movie! Female space aliens (played by Oriental women) use an invisible force field to trap Earthlings inside a country hospital (but why??). With Edward Judd, Yoko Tani.

June 21—MAN IN OUTER SPACE (1961) Still another ambiguous and forgettable title! Film is a Czech import, considered by some to be a comedy, about a man who is accidentally blasted into outer space, and when he returns, he finds it is now the 25th century. Film is difficult to find today. With Milos Kopecky, Vit Olmer. (*World Beyond* aired at 1:00pm this week, possibly due to a sporting event).

June 28—RED PLANET MARS (1952) Rerun.

July 5—THE FLYING SAUCER (1964) Italian film about a space alien who lands and kidnaps people; supposedly played for intentional laughs. With Alberto Sordi, Monica Vitti, Eleonora Rossi Drago. Another film that is difficult to find today.

July 12—QUEEN OF OUTER SPACE (1958) Rerun.

July 19—JOURNEY TO THE SEVENTH PLANET (1961) Kidnapped astronauts have their minds sapped by space aliens. With John Agar, Greta Thyssen, Carl Ottosen, Ann Smyrner. A U.S./Danish co-production.

July 26—IT CAME FROM OUTER SPACE (1953) Rerun.

August 2—THE TERROR (1963) On the Baltic Coast, a young Army military officer follows a beautiful ghost to a castle inhabited by an insane Baron. With Boris Karloff, Jack Nicholson (before he hit the big time), Sandra Knight, Dick Miller, Jonathan Haze. Directed by Roger Corman, who allegedly filmed this in two days! Others have disputed this legend, claiming that only the interior scenes with Karloff were shot in the 48-hour period.

August 9—ATTACK OF THE MUSHROOM PEOPLE (1963) Japanese sci-fi with shipwrecked tourists on a deserted Island encountering a deadly fungus that comes after them. With Akira Kubo.

August 16—BILLY THE KID VS. DRACULA (1966) The legendary outlaw decides to settle down and get married, but learns his fiancee's uncle is a vampire! Despite the title, the bloodsucker is never referred to on-screen as Dracula. Long regarded by horror fans as one of the worst films ever, its dismal reputation is due mostly to its absurd title. Otherwise, if you can accept the premise of vampires in the Old West, this film (while not good) is nowhere near as bad as many low-budget horror flicks. With John Carradine, Chuck Courtney, Virginia Christine, Bing Russell (Kurt's father in real life), Harry Carey Jr.

August 23—REVENGE OF THE CREATURE (1955) The first sequel to CREATURE FROM THE BLACK LAGOON finds the gill-man captured and put in a large aquarium. With John Agar, Lori Nelson, John Bromfield, Nestor Paiva. Film has attracted some renewed interest in later years as the film debut of Clint Eastwood, in an unbilled bit part as a lab technician (everyone has to start *somewhere!*).

August 30—WARNING FROM SPACE (1963?) Japanese-made sci-fi with space aliens who look like giant starfish coming to Earth to warn us of the dangers of the H-Bomb. The year this was made is unclear, as various sources list everything from 1956 to 1964.

September 6—GOG (1954) Rerun.

September 13—THE ATOMIC SUBMARINE (1959) Rerun. (With this week, the midnight repeat of *The World Beyond* was dropped by KPHO).

September 20—WORLD WITHOUT END (1956) Rerun.

September 27—CALTIKI, THE IMMORTAL MONSTER (1960) Rerun.

October 4—DOCTOR BLOOD'S COFFIN (1961) Another mad scientist holes up in an abandoned mine where he tries to resurrect the dead. With Kieron Moore, Hazel Court, Ian Hunter.

October 11—THE AMAZING COLOSSAL MAN (1957) Exposure to an atomic blast causes a soldier to grow into a giant! He also goes crazy and threatens the world! Produced and directed by Bert I. Gordon. With Glenn Langan, Cathy Downs, James Seay. The sequel to this, WAR OF THE COLOSSAL BEAST, had already been shown on *The World Beyond* twice.

October 18—DIARY OF A MADMAN (1963) A 19th-century French magistrate is possessed by an evil spirit called The Horla, which forces him to commit a variety of murders. With Vincent Price, Nancy Kovack, Ian Wolfe.

October 25—DR. TERROR'S HOUSE OF HORRORS (1965) Five short horror stories are dramatized, as mysterious doctor tells fortunes on a train one dark night. With Peter Cushing, Christopher Lee, Michael Gough, Donald Sutherland (before he found stardom). This was the first movie with horror film legend Christopher Lee to air on *The World Beyond.*

November 1—X-15 (1961) Pilots test experimental planes that may be able to fly into space. Not really science-fiction, though it has been so regarded by some. With Charles Bronson, Patricia Owens, Mary Tyler Moore (!), Kenneth Tobey, and narrated by James Stewart! Robert Dornan has an unbilled bit—he did some acting before becoming a reactionary right-wing U.S. Congressman in the 1990s.

November 8—HOUSE OF WAX (1953) A wax museum owner uses corpses to mold his figures. Near legendary horror movie, thanks to its 3-D effects that were used in theatres, and regarded as almost the definitive Vincent Price horror film. Also with Frank Lovejoy, Carolyn Jones, Charles Bronson (still billed as Charles Buchinsky in those days). This was a remake of MYSTERY OF THE WAX MUSEUM (1933), which would air on *The World Beyond* in 1984.

November 15—DESTINATION MOON (1950) Then-speculative dramatization of a flight to the moon. Produced by George Pal. With John Archer, Warner Anderson, Tom Powers. This film won the Academy Award for Best Special Effects.

November 22—THEY CAME FROM BEYOND SPACE (1967) Man is able to fight an invasion of space aliens because a steel plate in his head makes him immune to their evil powers. With Robert Hutton, Jennifer Jayne, Michael Gough.

November 29—INVADERS FROM SPACE (1964) Rerun.

December 6—WARNING FROM SPACE (1963?) Rerun.

December 13—RETURN OF THE FLY (1959) Rerun.

December 20—THEM! (1954) Rerun.

December 27—RED PLANET MARS (1953) Rerun.

1970

January 3—It appears *The World Beyond* was pre-empted by the American Bowl, a football game.

January 10—YOU'LL FIND OUT (1940) Musical comedy with bandleader Kay Kyser and his orchestra spending the night in an old dark house and foiling a plot to kill a young heiress. This was probably scheduled on *World Beyond* simply because of its supporting cast—Boris Karloff, Bela Lugosi, and Peter Lorre play the villains (this was the first Peter Lorre film to air on *The World Beyond*).

January 17—THE MONSTER THAT CHALLENGED THE WORLD (1957) Rerun.

On the following six Saturdays, *The World Beyond* was pre-empted by WAC Basketball games.

March 7—THE TERRORNAUTS (1967) Space aliens kidnap research scientists and take them to another planet. With Simon Oates, Zena Marshall. (This aired at 3:00pm following a PGA Golf Tournament).

March 14—THE UNEARTHLY STRANGER (1963) Rerun. (This aired at 3:00pm following a PGA Golf Tournament).

March 21—THE DAY THE EARTH FROZE (1964) Rerun. (This aired at 2:00pm)

March 28—THE BLACK SCORPION (1957) Rerun. (This aired at 1:30pm)

April 4—JOURNEY TO THE SEVENTH PLANET (1961) Rerun.

At this point, KPHO moved *The World Beyond* to a new time slot—Saturday mornings at 10:30am, where it stayed for the remaining 18 years of its run. This is significant because long term Arizona residents who grew up watching *The World Beyond* have fond memories of this slot. *The World Beyond*'s viewing audience clearly grew considerably after it moved into this position.

April 11—WAR OF THE SATELLITES (1958) Rerun.

April 18—BATTLE BEYOND THE SUN (1963) Rerun.

April 25—BEAST FROM 20,000 FATHOMS (1953) Rerun.

May 2—FRANKENSTEIN 1970 (1958) Rerun.

May 9—REPTILICUS (1962) A Danish film in which a giant prehistoric reptile attacks Copenhagen! With Carl Ottensen and Ann Smyrner, both of whom had also been in JOURNEY TO THE SEVENTH PLANET.

May 16—THE GIANT BEHEMOTH (1959) Rerun.

May 23—THE PHANTOM PLANET (1961) Rerun.

May 30—BEYOND THE TIME BARRIER (1959) Rerun.

June 6—QUEEN OF OUTER SPACE (1958) Rerun.

June 13—SON OF KONG (1933) Largely forgotten, and largely inferior, sequel to the legendary KING KONG (which never aired on *World Beyond*), with some of the same cast members. This time, they return to the Island and discover a cute baby Kong who is far less menacing than his daddy was. The special effects are good, but the film clearly was rushed into production and release the same year as the original Kong. With Robert Armstrong, Helen Mack.

June 20—Some confusion here. KPHO's paid ad in The Phoenix Gazette newspaper lists VILLAGE OF THE GIANTS, a ridiculous Bert I. Gordon-produced film about giant teenagers which starred Tommy Kirk, Ronny Howard (!), and a young Beau Bridges. However, the Gazette's printed tv schedule for the day lists MEXICAN SPITFIRE SEES A GHOST, a 1942 haunted house comedy with Lupe Velez, Leon Errol, and Mantan Moreland. We do not know which film actually aired on *World Beyond*, though it was probably VILLAGE OF THE GIANTS, since the tv schedule would have been submitted to the newspaper long before the ad.

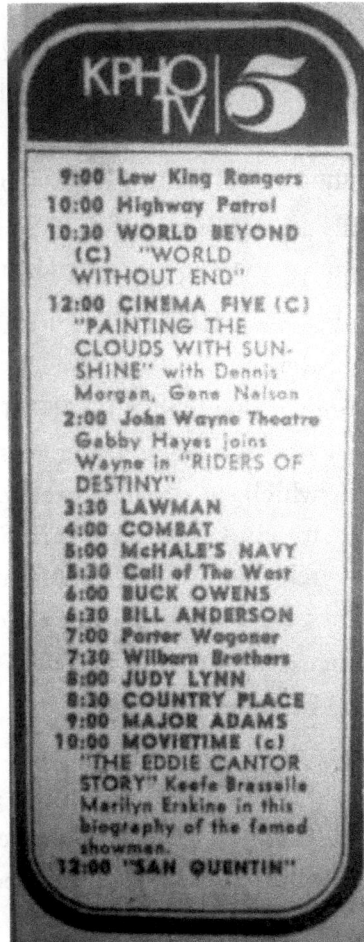

June 27—ABBOTT AND COSTELLO MEET FRANKENSTEIN (1948) Regarded as one of the famed comedy team's best films, and it was also the last appearance of the legendary "Universal monsters."

Oddly, it was the first film with the Frankenstein monster and also the Wolf Man to air on *The World Beyond*. Also with Lon Chaney Jr., Bela Lugosi (as Dracula), Glenn Strange, and an unbilled voice bit by Vincent Price.

July 4—SATELLITE IN THE SKY (1956) Rerun.

July 11—INVASION OF THE STAR CREATURES (1961) Rerun.

July 18—WORLD WITHOUT END (1956) Rerun.

July 25—CREATURE FROM THE BLACK LAGOON (1954) Rerun.

August 1—CASTLE OF BLOOD (1963) Italian film about a man who accepts a wager to spend a night in a haunted castle from which no one has ever returned alive. This one is quite good of its kind. With Barbara Steele and Georges Riviere. It was also released under the alternate title of CASTLE OF TERROR, and was rerun on *The World Beyond* under that name in 1978.

August 8—ATTACK OF THE MUSHROOM PEOPLE (1963) Rerun.

August 15—CASTLE OF THE LIVING DEAD (1963) In Medieval times, a motley group of people travel to the eerie castle of Count Drago, where terror awaits them! A low-budget Italian film with Christopher Lee and Donald Sutherland (before he hit stardom).

August 22—ZOMBIES ON BROADWAY (1945) The forgotten comedy team of Wally Brown and Alan Carney star as press agents who uncover a plot to create zombies to appear at a nightclub. Also with Bela Lugosi, Sheldon Leonard.

August 29—PHANTOM OF THE RUE MORGUE (1954) A zoologist hypnotizes an ape to commit a series of grisly murders in the loose adaptation of Edgar Allen Poe's much filmed story MURDERS IN THE RUE MORGUE. With Karl Malden, Patricia Medina, Claude Dauphin, Steve Forrest, Merv Griffin (!).

September 5—THE COLOSSUS OF NEW YORK (1958) Mad scientist transplants the brain of his dead son into a giant robot! With John Baragrey, Mala Powers, Otto Kruger, Robert Hutton, Ross Martin.

September 12—IT CAME FROM OUTER SPACE (1953) Rerun.

September 19—THE CREATURE WALKS AMONG US (1956) Rerun.

September 26—RETURN OF THE FLY (1959) Rerun.

October 3—THE UNEARTHLY STRANGER (1963) Rerun.

October 10—ABBOTT AND COSTELLO GO TO MARS (1953) Rerun.

October 17—THE DEMON PLANET (1965) Astronauts land on a strange planet in search of a previous spaceship that disappeared. Soon, an evil force starts to control their minds! An Italian film, in which director Mario Bava managed to take a routine science-fiction script of the era and, through sheer imagery and atmosphere alone, turn it into a rather eerie film. Not bad at all. With Barry Sullivan, Norma Bengell. This had an alternate title of PLANET OF THE VAMPIRES.

October 24—THEY CAME FROM BEYOND SPACE (1967) Rerun.

October 31—MASTER OF THE WORLD (1961) A 19th century madman attempts to use his giant flying airship to stop war in the world. Quite good, with Vincent Price, Charles Bronson, Henry Hull, Richard Harrison.

November 7—GODZILLA VS. THE THING (1964) The Thing is actually Mothra, a giant moth who had appeared in her own film two years before (and which was oddly never shown on *The World Beyond*). Godzilla was still a sinister character at this point in the famous series of Japanese monster movies, and Mothra tries to stop him from destroying Tokyo. In more recent years, the distribution rights to this film changed hands, and its new owners retitled it simply GODZILLA VS. MOTHRA for all video, DVD, and tv prints.

November 14—*The World Beyond* seems to have been pre-empted by a Jr. High School Football Championship between Arizona teams Safford and Glendale's Unit 6.

November 21—PLANET OF BLOOD (1966) Astronauts find and capture a female blood-drinking alien and bring her to Earth. Directed by Curtis Harrington with some footage of spaceships from a Russian film (PLANETA BURG) edited in patchwork style. This had an alternate title of QUEEN OF BLOOD. With John Saxon, Basil Rathbone, Dennis Hopper, Florence Marly, Forrest J. Ackerman.

November 28—THE DAY THE EARTH FROZE (1959) Rerun.

December 5—PORT SINISTER (1952) An earthquake unleashes an invasion of giant crabs! Very low-budget opus, very difficult to find today. With James Warren, William Schallert, House Peters Jr. (son of a once-famous silent film star).

December 12—BEYOND THE TIME BARRIER (1959) Rerun.

December 19—X THE UNKNOWN (1956) Rerun.

December 26—THE THREE STOOGES IN ORBIT (1962) The famed comedy team foils a plot by Martians to steal a professor's ray gun invention. Also with Carol Christensen, Emil Sitka, Nestor Paiva.

1971

January 2—SATELLITE IN THE SKY (1956) Rerun.

January 9—MISSILE TO THE MOON (1959) More astronauts meet more sexy alien women in this extremely low-budget opus. Some people regard this as so bad it's fun! Directed by Richard Cunha. With Richard Travis, Cathy Downs.

January 16—MAN IN OUTER SPACE (1961) Rerun.

January 23—SPACE MONSTER (1964) Rerun.

January 30—BATTLE BEYOND THE SUN (1963) Rerun.

February 6—JOURNEY TO THE SEVENTH PLANET (1961) Rerun.

February 13—FRANKENSTEIN CONQUERS THE WORLD (1966) Japanese film with a giant-size caveman-like creature battling a dinosaur called Baragon (who later returned in some other Japanese monster movies). One of the more absurd creature features from the land of the rising sun, with no real relation to Frankenstein. With Nick Adams, Seuko Tagami.

February 20—VOYAGE TO THE END OF THE UNIVERSE (1964) A Czechoslovakia import, with a group of ambiguous space travelers (from another planet, maybe?) searching for a new, friendlier planet to live on. With Francis Smolen.

February 27—THE MOST DANGEROUS MAN ALIVE (1961) Escaped convict gets super powers from an atomic bomb blast, seeks revenge on those who put him in prison. With Ron Randell, Debra Paget, Morris Ankrum.

March 6—DIE, MONSTER, DIE! (1965) Young man goes to England to visit his fiancée's family, discovers her father is a mad scientist experimenting with a radioactive meteorite. With Boris Karloff, Nick Adams, Patrick Magee, Suzy Parker.

March 13—ABBOTT AND COSTELLO MEET DR. JEKYLL AND MR. HYDE (1953) The famed comedy team meets more monsters. Also with Boris Karloff, Craig Stevens, Helen Westcott, Reginald Denny.

March 20—THE PHANTOM PLANET (1961) Rerun.

March 27—REPTILICUS (1961) Rerun.

April 3—THE GIANT BEHEMOTH (1959) Rerun.

April 10—MONSTER FROM THE SURF (1965) Teenage beach-partyers appear to be murdered one by one, but the killer turns out to be a crazy oceanographer. With Jon Hall (who also directed).

April 17—DAGORA, THE SPACE MONSTER (1964) Rerun.

April 24—THE THING (1951) Rerun.

May 1—THE TERRORNAUTS (1967) Rerun.

May 8—SON OF KONG (1933) Rerun.

May 15—ABBOTT AND COSTELLO MEET THE MUMMY (1955) The title pretty much says it all, as the famed comedy team finds adventures in Egypt. Also with Marie Windsor.

May 22—THE DAY THE EARTH STOOD STILL (1951) Regarded to this day as one of the most significant science fiction movies ever made, as space alien lands on Earth to warn people they need to change their ways, but finds himself on the run from the military. Directed by Robert Wise. With Michael Rennie, Patricia Neal, Hugh Marlowe, Sam Jaffe, Frances Bavier.

May 29—BEYOND THE TIME BARRIER (1959) Rerun, which began at 10:00am just for this week.

June 5—WORLD WITHOUT END (1956) Rerun.

June 12—ZOTZ! (1962) A professor finds a coin that gives him magic powers in this comedy produced and directed by William Castle. With Tom Poston, Fred Clark, Jim Backus, Cecil Kellaway.

June 19—INVASION OF THE STAR CREATURES (1961) Rerun.

June 26—THE THREE STOOGES MEET HERCULES (1961) They travel back in time to ancient Greece where they are kidnapped as galley slaves, meet a giant cyclops, and eventually find Hercules himself. Also with Emil Sitka.

July 3—CREATURE FROM THE BLACK LAGOON (1954) Rerun.

July 10—X-THE MAN WITH THE X-RAY EYES (1963) Rerun.

July 17—REVENGE OF THE CREATURE (1955) Rerun.

July 24—IT CAME FROM OUTER SPACE (1953) Rerun.

July 31—ATTACK OF THE MUSHROOM PEOPLE (1963) Rerun.

August 7—THE CREATURE WALKS AMONG US (1956) Rerun.

August 14—THE UNEARTHLY STRANGER (1964) Rerun.

August 21—ATTACK OF THE ROBOTS (1962) Interpol secret agent named Lemmy Caution saves the world from robot-men who have been programmed to assassinate world leaders. This Italian film was reportedly an entry in a low-budget series of Lemmy Caution movies in its home country. Directed by Jesus Franco. With Eddie Constantine, Fernando Rey.

August 28—THE COLOSSUS OF NEW YORK (1958) Rerun.

September 4—A bit of confusion here with contradictory listings again. KPHO's paid ad in The Arizona Republic that day lists THE THREE STOOGES IN ORBIT, while the paper's regular tv listings have CURSE OF THE SWAMP CREATURE, a 1966 opus with John Agar and Francine York. We do not know which one was actually shown.

September 11—IN THE YEAR 2889 (1967) Following a nuclear holocaust, survivors hide out from mutants. Directed by Larry Buchanan. With Paul Petersen, Charla Doherty.

September 18—THEY CAME FROM BEYOND SPACE (1967) Rerun.

September 25—FRANCIS IN THE HAUNTED HOUSE (1956) This was the last entry in the "Francis the Talking Mule" series, with a title that pretty much explains it all. This one had a human cast change, with Mickey Rooney taking over the role of Francis' buddy from Donald O'Connor, and Paul Frees replacing Chill Wills as the voice of Francis.

October 2—ABBOTT AND COSTELLO MEET THE KILLER, BORIS KARLOFF (1949) This time, A & C play amateur detectives who try to solve a series of murders. Karloff plays an evil hypnotist-swami. The "star name in the title" gimmick failed to catch on. Also with Roland Winters.

October 9—Another case of conflicting data here. The Arizona Republic's regular tv listings have a rerun of SATELLITE IN THE SKY, while KPHO's paid ad in the newspaper lists VALLEY OF THE DRAGONS, a 1961 film about earthmen landing on a comet with monsters. We do not know which movie was aired on *World Beyond*.

October 16—MUTINY IN OUTER SPACE (1964) A deadly growing fungus menaces astronauts on a space ship. First showing of a movie that was aired much too often on *The World Beyond*. With William Leslie, Dolores Faith, Glenn Langan, Richard Garland, Harold Lloyd Jr. (!).

October 23—THE HUMAN DUPLICATORS (1965) A very tall space alien invades Earth by creating duplicates of important Earthlings who will do his bidding. Absolutely awful, but almost fun because of its bizarre cast! With George Nader, Barbara Nichols, Richard Kiel, Dolores Faith, George Macready, Richard Arlen, Hugh Beaumont.

October 30—THE UNDERWATER CITY (1962) Scientists try to build the first city under the sea. It tries to be more "thoughtful" than most low-grade sci-fi, but its small budget really hurts it. Produced by Alex Gordon. With William Lundigan, Julie Adams.

November 6—THE BLOB (1958) Near-legendary low-budget film about a mass of goo that devours people and grows and grows (and grows.....). It's many fans consider the film to be a laugh riot, but like THE FLY, I found it surprisingly grisly for this sort of thing. With Steve McQueen (before he hit stardom), Aneta Corsaut, Olin Howlin. It was remade with a big budget in 1988!

November 13—DINOSAURUS! (1960) Two dinosaurs and a caveman are accidentally unearthed in modern times. They come back to life and run amok! Quite good special effects help keep things going. With Ward Ramsey, Paul Lukather.

November 20—THE 4D MAN (1959) Mad scientist learns how to make himself be able to pass through solid matter, and goes on a spree of robbery and murder. With Robert Lansing, Lee Meriweather, Patty Duke (!).

November 27—MUTINY IN OUTER SPACE (1964) Rerun, surprisingly soon after its October 16 premiere on *The World Beyond*.

December 4—MAN IN OUTER SPACE (1961) Rerun, which aired at 11:00am following the broadcast of a Gompers Parade.

December 11—GODZILLA VS. THE THING (1964) Rerun.

December 18—SPACE MONSTER (1964) Rerun.

December 25—*The World Beyond* was pre-empted by two Christmas specials—"Unto Us A Child Is Born", and a Christmas Mass for Shut-Ins.

1972

January 1—THE BLACK SCORPION (1957) Rerun.

January 8—DAGORA, THE SPACE MONSTER (1963) Rerun.

January 15—REPTILICUS (1961) Rerun.

January 22—JOURNEY TO THE SEVENTH PLANET (1961) Rerun.

January 29—THE PHANTOM PLANET (1961) Rerun.

February 5—THE THING (1951) Rerun.

February 12—RODAN (1956) Giant pterodactyl hatches from an egg found in a mine, and causes destruction! Japanese movie was the debut of Rodan, who returned in later Godzilla movies.

February 19—GODZILLA'S REVENGE (1967) Toho Studios, the producers of the Godzilla movies, started experimenting with this one. Bullied latchkey boy escapes his troubles by dreaming of visiting Monster Island and meeting Godzilla's wise-cracking son

Minya! Needless to say, the title is very misleading. Stock footage of monster battles from other Godzilla movies is edited in also. Since the monsters are not real—only dreams—in this one, many viewers were disappointed with it (though it does have its defenders) and Toho went back to producing "regular" monster movies after this.

February 26—VOYAGE TO THE END OF THE UNIVERSE (1964) Rerun.

March 4—DIE, MONSTER, DIE! (1965) Rerun.

March 11—BEYOND THE TIME BARRIER (1959) Rerun.

March 18—INVASION OF THE STAR CREATURES (1961) Rerun.

March 25—More newspaper confusion. KPHO's paid ad lists a rerun of THE AMAZING TRANSPARENT MAN, while The Phoenix Gazette's tv listings show a rerun of REVENGE OF THE CREATURE. Again, we do not know which one was correct.

April 1—X THE UNKNOWN (1956) Rerun.

April 8—ABBOTT AND COSTELLO GO TO MARS (1953) Rerun.

April 15—X-THE MAN WITH THE X-RAY EYES (1963) Rerun.

April 22—FRANKENSTEIN CONQUERS THE WORLD (1966) Rerun.

April 29—THE COLOSSUS OF NEW YORK (1958) Rerun.

May 6—MAN-EATER OF HYDRA (1966) Spanish import, with a motley group of people stranded on an Island where a mad botanist is breeding omnivorous, bloodthirsty plants! Directed by Mel Welles. With Cameron Mitchell, Elisa Montes, George Martin.

May 13—CURSE OF THE SWAMP CREATURE (1966) Possible rerun from September 4, 1971. A mad scientist creates reptile monsters in a swamp. Directed by Larry Buchanan. With John Agar, Francine York.

May 20—THEY CAME FROM BEYOND SPACE (1967) Rerun.

May 27—CREATURE FROM THE BLACK LAGOON (1954) Rerun.

June 3—NIGHT STAR, GODDESS OF ELECTRA (1965) In Ancient Rome, a mad magician resurrects dead soldiers to do his evil bidding, while a brawny hero tries to stop him. Italian film was also released under the name WAR OF THE ZOMBIES, and is difficult to find today. With John Drew Barrymore, Susy Anderson.

June 10—ATTACK OF THE ROBOTS (1962) Rerun.

June 17—IN THE YEAR 2889 (1965) Rerun.

June 24—THE CREATURE WALKS AMONG US (1956) Rerun.

July 1—IT CAME FROM OUTER SPACE (1953) Rerun.

July 8—VALLEY OF THE DRAGONS (1961) Possible rerun from October 9, 1971. Two earthmen find themselves stranded on a comet inhabited by monsters. With Cesare Danova, Sean McClory.

July 15—REVENGE OF THE CREATURE (1955) Rerun.

July 22—TEENAGERS FROM OUTER SPACE (1959) Rerun.

July 29—PLANET OF BLOOD (1966) Rerun.

August 5—MUTINY IN OUTER SPACE (1964) Rerun.

August 12—CREATURE OF DESTRUCTION (1967) After a hypnotist predicts murders will occur at a country club, a monster rises from a lake and goes on a rampage. Directed by Larry Buchanan. With Les Tremayne, Pat Delaney. *The World Beyond* began at 10:00am this week because KPHO had to shift their schedule around to air the Westchester Golf Classic tournament at 1:00pm).

August 19—IT CAME FROM OUTER SPACE (1953) Rerun, only about six weeks after its previous showing.

August 26—IT'S ALIVE! (1968) A madman kidnaps people to feed to his pet monster who lives in a cave. Directed by Larry Buchanan. With Tommy Kirk, Shirley Bonn. Readers of this book should not confuse this film with a 1974 movie of the same name about a killer baby!

September 2—THE 4D MAN (1959) Rerun.

September 9—GODZILLA VS. THE THING (1964) Rerun.

September 16—DIE, MONSTER, DIE! (1965) Rerun.

September 23—THE TERRORNAUTS (1967) Rerun.

September 30—GODZILLA'S REVENGE (1967) Rerun.

October 7—SON OF DRACULA (1943) Count Dracula (not his son) puts the bite on people at a Louisiana plantation while disguised as "Count Alucard". Universal-produced film has good atmosphere and is underrated. Directed by Robert Siodmak. With Lon Chaney Jr. (his only time portraying Dracula), Louise Allbritton, Evelyn Ankers, Robert Paige.

October 14—FRANKENSTEIN MEETS THE SPACE MONSTER (1965) Very low-budget nonsense about a man-turned-robot (the "Frankenstein" of the title) battling a space monster. Regarded as one of the all-time worst! With James Karen, David Kerman, Nancy Marshall.

October 21—THE WOLF MAN (1941) One of the all-time classic monster movies, with Lon Chaney Jr. beginning the role he would remain most famous for—Lawrence Talbot, a man who becomes a werewolf and wishes to die because of it. Written by Curt Siodmak. Also with Claude Rains, Warren William, Ralph Bellamy, Bela Lugosi, Evelyn Ankers, Maria Ouspenskaya. Patric Knowles.

October 28—RODAN (1956) Rerun.

November 4—CURSE OF THE MUMMY'S TOMB (1964) A showman brings a mummy to England for display, but naturally it comes to life and goes on a rampage. Produced by Hammer Films, and not one of their better efforts. With Terence Morgan, Fred Clark (!), Ronald Howard. I have a personal memory of seeing this movie as a young adolescent, and being excited by a sequence where the camera just lingers on the anatomy of a belly dancer for a prolonged period.....

November 11—THE NAVY VS. THE NIGHT MONSTERS (1966) Moving, killer trees attack a navy base! The bizarre cast has given this rock-bottom cheapie a small following in later years. With Mamie Van Doren, Anthony Eisley, Bobby Van.

November 18—WOMEN OF THE PREHISTORIC PLANET (1965) More astronauts explore a mysterious planet with monsters and alien humans—and a twist ending. With Wendell Corey, John Agar, Keith Larsen, Merry Anders, Adam Roarke, Stuart Margolin.

November 25—CURSE OF THE CAT PEOPLE (1944) Very strange sequel to the classic THE CAT PEOPLE (1942), which would be shown on *The World Beyond* many years later in 1985. Little girl is haunted by visions of her father's evil first wife (from the first film). With Simone Simon, Kent Smith.

December 2—VOYAGE TO THE END OF THE UNIVERSE (1964) Rerun.

December 9—THE UNDERWATER CITY (1962) Rerun. Aired at 11:00am this week because of a Gompers parade in Phoenix.

December 16—MAN-EATER OF HYDRA (1966) Rerun.

December 23—DRACULA'S DAUGHTER (1936) Lesser-remembered direct sequel to the 1931 DRACULA, with Professor Van Helsing pursuing the title character in London. With Otto Kruger, Gloria Holden, Edward Van Sloan, Marguerite Churchill, Hedda Hopper.

December 30—THE BRIDE OF FRANKENSTEIN (1935) Considered by film historians to be one of the greatest horror films ever made, as the undying Monster roams the countryside, while Dr. Frankenstein tries to create a female creature. Everything works in this classic, and its scenes have been imitated and parodied countless times. With Boris Karloff, Colin Clive, Elsa Lanchester, Valerie Hobson, Ernest Thesiger, O.P. Heggie, Una O'Connor, Dwight Frye, John Carradine. Directed by James Whale.

1973

January 6—THE THING (1951) Rerun.

January 13—FIRST MAN INTO SPACE (1959) An astronaut returns to Earth as monster who drinks blood. With Marshall Thompson, Marla Landi.

January 20—ATTACK FROM SPACE (1964) Rerun.

January 27—CREATION OF THE HUMANOIDS (1962) Rerun.

February 3—THE HUMAN DUPLICATORS (1965) Rerun.

February 10—PLANET ON THE PROWL (1965) Italian film with astronauts trying to stop the orbit of another planet which is causing major natural disasters on Earth. Really boring. With Jack Stuart, Amber Collins.

February 17—THE ATOMIC SUBMARINE (1959) Rerun.

February 24—THE TERRORNAUTS (1967) Rerun.

March 3—TERROR BENEATH THE SEA (1965) Mad scientist in undersea laboratory turns people into monsters! Japanese-made sci-fi, with Mike Daneen, Sinichi Chiba (later known as Sonny Chiba in martial-arts movies). Another film that is not easy to locate today.

March 10—ABBOTT AND COSTELLO MEET THE INVISIBLE MAN (1951) The title pretty much says it all again, as the comedy team gets involved with an invisible prize-fighter. Also with Arthur Franz, Sheldon Leonard.

March 17—THE COLOSSUS OF NEW YORK (1958) Rerun.

March 24—ABBOTT AND COSTELLO GO TO MARS (1953) Rerun.

March 31—THEY CAME FROM BEYOND SPACE (1967) Rerun.

April 7—CREATURE FROM THE BLACK LAGOON (1954) Rerun.

April 14—ABBOTT AND COSTELLO MEET THE KILLER, BORIS KARLOFF (1949) Rerun.

April 21—VALLEY OF THE DRAGONS (1961) Rerun.

April 28—DINOSAURUS! (1960) Rerun.

May 5—WOMEN OF THE PREHISTORIC PLANET (1965) Rerun.

May 12—THE NAVY VS. THE NIGHT MONSTERS (1966) Rerun.

May 19—THE 4D MAN (1959) Rerun.

May 26—THE FABULOUS BARON MUNCHAUSEN (1961) Another Czech import, dramatizing the escapades of one of the world's best known folklore figures. With Milos Kopecky.

June 2—ATTACK OF THE ROBOTS (1962) Rerun.

June 9—THE DEMON PLANET (1965) Rerun.

June 16—GODZILLA VS. THE THING (1964) Rerun.

June 23—DIE, MONSTER, DIE! (1965) Rerun.

June 30—IN THE YEAR 2889 (1965) Rerun.

July 7—CURSE OF THE SWAMP CREATURE (1966) Rerun.

July 14—WAR OF THE GARGANTUAS (1967) Another Japanese monster movie, with two giant hairy critters battling it out, one good and one evil! With Russ Tamblyn.

July 21—GODZILLA'S REVENGE (1967) Rerun.

July 28—RODAN (1956) Rerun.

August 4—MONSTER ZERO (1968) Japanese-made sci-fi in which aliens who control Ghidrah the Three-Headed Monster threaten the Earth, until Godzilla and Rodan step in. A little more elaborate than most of the Toho-produced monster movies. With Nick Adams. A change in distributor rights in America in recent years resulted in a title change to GODZILLA VS. MONSTER ZERO, and prints with this title seem to be the only ones currently available.

August 11—BLACK FRIDAY (1940) A scientist transplants the brain of a dead gangster into another man, turning him into a dangerous criminal. With Boris Karloff, Bela Lugosi, Stanley Ridges. Film was, and remains to this day, controversial to horror fans because Lugosi, despite second billing, has only a small role and no scenes with Karloff!

August 18—FRANKENSTEIN (1931) A landmark among horror films, debuting the Universal Studio's version of the Frankenstein monster, as portrayed by Boris Karloff. Not much can be said about this film that has not already been said. Also with Colin Clive, Mae Clarke, John Boles, Edward Van Sloan, Frederick Kerr, Dwight Frye, Lionel Belmore, Marilyn Harris. Directed by James Whale.

August 25—DRACULA (1931) Arguably the most legendary horror film ever made; the film that made Bela Lugosi a star while simultaneously ruining his life and career. Most cinematic vampire clichés had their origins here, and as with FRANKENSTEIN, there isn't much to say about this film that hasn't already been said. Directed by Tod Browning. Also with Helen Chandler, David Manners, Edward Van Sloan, Dwight Frye. Incidentally, no film was ever aired on *The World Beyond* older than 1931.

September 1—FRANKENSTEIN MEETS THE SPACE MONSTER (1965) Rerun. Imagine *The World Beyond* airing this following the showings of two of the greatest monster movies ever made!

September 8—THE TERRORNAUTS (1967) Rerun.

September 15—MUTINY IN OUTER SPACE (1964) Rerun.

September 22—THE BLOB (1958) Rerun.

September 29—THE GORGON (1964) A very well-done Hammer chiller, wherein the snake-headed demon from Greek mythology who can turn people to stone terrorizes a small village. With Peter Cushing, Christopher Lee, Barbara Shelley.

October 6—INVASION OF THE NEPTUNE MEN (1964) Rerun.

October 13—INVADERS FROM SPACE (1964) Rerun.

October 20—IT'S ALIVE! (1968) Rerun.

October 27—CURSE OF THE MUMMY'S TOMB (1964) Rerun.

November 3—TERROR BENEATH THE SEA (1965) Rerun.

November 10—PLANET ON THE PROWL (1965) Rerun.

November 17—FRANKENSTEIN CONQUERS THE WORLD (1966) Rerun.

November 24—ATOMIC RULERS OF THE WORLD (1964) Rerun.

December 1—VOYAGE TO THE END OF THE UNIVERSE (1964) Rerun.

December 8—THE UNDERWATER CITY (1962) Rerun.

December 15—PRINCE OF SPACE (1962) Rerun.

December 22—THE THING (1951) Rerun.

December 29—PLANETS AGAINST US (1961) Rerun.

1974

January 5—VALLEY OF THE DRAGONS (1961) Rerun.

January 12—THE ATOMIC SUBMARINE (1959) Rerun.

January 19—ATTACK FROM SPACE (1964) Rerun.

January 26—FIRST MAN INTO SPACE (1959) Rerun.

February 2—THE DEMON PLANET (1965) Rerun.

February 9—DINOSAURUS! (1960) Rerun.

February 16—THE 4D MAN (1959) Rerun.

February 23—DR. GOLDFOOT AND THE BIKINI MACHINE (1966) Comedy about a mad scientist who tries to take over the world using his army of beautiful female robots. With Vincent Price, Frankie Avalon, Susan Hart, Dwayne Hickman, Fred Clark. Price always claimed to be very choosy about what films he did, and he was occasionally confronted by interviewers in later years who wanted to know what he saw in this silly twaddle.

March 2—RODAN (1956) Rerun.

March 9—THEY CAME FROM BEYOND SPACE (1967) Rerun.

March 16—ATTACK OF THE ROBOTS (1962) Rerun.

March 23—THE INVISIBLE MAN RETURNS (1940) First sequel to Universal's THE INVISIBLE MAN, with a man convicted of murder using invisibility to escape from prison and clear himself. With Vincent Price, Sir Cedric Hardwicke, Cecil Kellaway. Regarded today as Price's first "real" horror film, though he would not become synonymous with the genre for nearly 20 years yet. Screenplay by Lester Cole and Curt Siodmak.

March 30—SON OF DRACULA (1943) Rerun.

April 6—THE GORGON (1964) Rerun.

April 13—FRANKENSTEIN MEETS THE WOLF MAN (1943) Universal's two biggest monster "stars" meet and battle it out at the climax. This film has been the subject of much controversy over the years, due to its disastrous casting of Bela Lugosi as the Frankenstein monster, which resulted in much post-production editing (to eliminate the Monster's speaking lines) before release, and the casting of a stunt double for Lugosi in many of the scenes. Still, it is a film that is very hard to dislike! Also with Lon Chaney Jr., Ilona Massey, Patric Knowles, Maria Ouspenskaya, Lionel Atwill, Dwight Frye. Screenplay by Curt Siodmak.

April 20—MUTINY IN OUTER SPACE (1965) Rerun.

April 27—VISIT TO A SMALL PLANET (1960) Loose adaptation of a Gore Vidal play about a comic space alien who comes to Earth to observe man's strange ways, and wreaks slapstick havoc. With Jerry Lewis, Joan Blackman, Fred Clark, Earl Holliman, Gale Gordon, Jerome Cowan. Needless to say, this was the only time a Jerry Lewis movie was shown on *The World Beyond*!

May 4—DIE, MONSTER, DIE! (1965) Rerun.

May 11—CURSE OF THE FLY (1966) Mad scientist conducts the usual ghastly experiments. Officially the final sequel to THE FLY (1958), but the giant human fly does not appear in this one except in a photo, and the film is slow and talky. With Brian Donlevy, Carole Gray.

May 18—THE HUMAN DUPLICATORS (1964) Rerun.

May 25—WAR-GODS OF THE DEEP (1965) A crew of men, searching for a missing girl, discover an underwater city ruled by a mad sea captain and his aquatic "gill men". Officially based on Edgar Allen Poe's poem "City in the Sea", but there is little resemblance. With Vincent Price, Tab Hunter, Susan Hart, David Tomlinson. Last film for director Jacques Tourneur.

June 1—THE MUMMY (1932) The original Universal film, with Boris Karloff as an evil Egyptian high priest who returns to life to seek the reincarnation of his lost love. Different from all later "mummy" films in that the title character speaks and looks relatively human instead of existing as a shambling, bandaged zombie. Also with Zita Johann, David Manners, Edward Van Sloan, Bramwell Fletcher.

June 8—FRANKENSTEIN MEETS THE SPACE MONSTER (1965) Rerun.

June 15—NIGHT MONSTER (1942) Rerun.

June 22—THE COLOSSUS OF NEW YORK (1958) Rerun.

June 29—BLACK FRIDAY (1940) Rerun.

July 6—THE INVISIBLE MAN (1933) Rerun.

July 13—THE WOLF MAN (1941) Rerun.

July 20—ATOMIC RULERS OF THE WORLD (1964) Rerun.

July 27—INVASION OF THE NEPTUNE MEN (1964) Rerun.

August 3—DRACULA'S DAUGHTER (1936) Rerun.

August 10—INVADERS FROM SPACE (1964) Rerun.

August 17—VOYAGE TO THE END OF THE UNIVERSE (1964) Rerun.

August 24—THE TERRORNAUTS (1966) Rerun.

August 31—PRINCE OF SPACE (1962) Rerun.

September 7—THE UNDERWATER CITY (1962) Rerun.

September 14—PLANETS AGAINST US (1961) Rerun.

September 21—THE BLOB (1958) Rerun.

September 28—THE EVIL BRAIN FROM OUTER SPACE (1964) Rerun.

October 5—DINOSAURUS! (1960) Rerun.

October 12—PLANET OF BLOOD (1966) Rerun.

October 19—CURSE OF THE MUMMY'S TOMB (1964) Rerun.

October 26—PLANET ON THE PROWL (1965) Rerun.

November 2—TERROR BENEATH THE SEA (1965) Rerun.

November 9—THE NAVY VS. THE NIGHT MONSTERS (1966) Rerun.

November 16—MAN-EATER OF HYDRA (1966) Rerun.

November 23—PRINCE OF SPACE (1962) Rerun.

November 30—MUTINY IN OUTER SPACE (1965) Rerun.

December 7—THE ATOMIC SUBMARINE (1959) Rerun.

December 14—ATTACK FROM SPACE (1964) Rerun.

December 21—ATTACK OF THE ROBOTS (1962) Rerun.

December 28—SLAUGHTER OF THE VAMPIRES (1962) Rerun.

1975

January 4—THE GORGON (1964) Rerun.

January 11—THE 4D MAN (1959) Rerun.

For three Saturdays in a row, *The World Beyond* was pre-empted by College Basketball games, preceded by an episode of "The Rifleman" tv series.

February 8—THE INCREDIBLE SHRINKING MAN (1957) After being exposed to a strange mist, a man shrinks away to nothing. Unusually grim for 1950s sci-fi, and this is regarded as a classic of its era. With Grant Williams, Randy Stuart, Paul Langton, William Schallert. Screenplay was by Richard Matheson. Produced by Albert Zugsmith and directed by Jack Arnold—this is the only film by either man that anyone ever says anything good about.

February 15—GORGO (1961) A dinosaur-like monster stomps on London while looking for its child! More fun than usual for a giant monster movie, with scenes of the beast destroying Westminster Abbey and Big Ben! With Bill Travers (who later went on to BORN FREE), William Sylvester (who later went on to 2001:A SPACE ODYSSEY!).

February 22—*The World Beyond* is once again pre-empted by a College Basketball game and a rerun of "The Rifleman" series.

March 1—THE MYSTERIANS (1957) Japanese-made sci-fi with space aliens invading Earth accompanied by a giant robot!

March 8—ATOMIC RULERS OF THE WORLD (1964) Rerun.

March 15—IT'S ALIVE! (1968) Rerun.

March 22—INVASION OF THE NEPTUNE MEN (1964) Rerun.

March 29—VALLEY OF THE DRAGONS (1961) Rerun.

April 5—CURSE OF THE FLY (1966) Rerun.

April 12—INVADERS FROM SPACE (1964) Rerun.

April 19—PLANETS AGAINST US (1961) Rerun.

April 26—FRANKENSTEIN (1931) Rerun.

May 3—WOMEN OF THE PREHISTORIC PLANET (1965) Rerun.

May 10—GODZILLA VS. THE THING (1964) Rerun.

May 17—THE HUMAN DUPLICATORS (1964) Rerun.

May 24—FRANKENSTEIN MEETS THE SPACE MONSTER (1965) Rerun.

May 31—MUTINY IN OUTER SPACE (1964) Rerun.

June 7—FRANKENSTEIN MEETS THE WOLF MAN (1943) Rerun.

June 14—THE MUMMY (1932) Rerun.

June 21—THE INVISIBLE MAN RETURNS (1940) Rerun.

June 28—FIRST MAN INTO SPACE (1959) Rerun.

July 5—THE BLOB (1958) Rerun.

July 12—DIE, MONSTER, DIE! (1965) Rerun.

July 19—THE DEMON PLANET (1965) Rerun.

July 26—PRINCE OF SPACE (1962) Rerun.

August 2—THE EVIL BRAIN FROM OUTER SPACE (1964) Rerun.

August 9—THE ATOMIC SUBMARINE (1959) Rerun.

August 16—THE WOLF MAN (1941) Rerun.

August 23—VOYAGE TO THE END OF THE UNIVERSE (1964) Rerun.

August 30—ATTACK FROM SPACE (1964) Rerun.

September 6—MAN-EATER OF HYDRA (1966) Rerun.

September 13—CREATION OF THE HUMANOIDS (1962) Rerun.

September 20—CURSE OF THE SWAMP CREATURE (1966) Rerun.

September 27—SLAUGHTER OF THE VAMPIRES (1962) Rerun.

October 4—DINOSAURUS! (1960) Rerun.

October 11—FROM THE EARTH TO THE MOON (1958) Loose film version of a Jules Verne story about a 19th century rocket to the moon. Unlike most 1950s sci-fi, this was a "big" studio production with an oddball cast of veteran actors. With Joseph Cotten, George Sanders, Debra Paget, Don Dubbins, Patric Knowles, Henry Daniell. KPHO clearly had to do considerable editing to fit this into *The World Beyond*'s 90-minute time slot that included commercials! It would later be rerun on KPHO's two-hour *Action Theatre*.

October 18—COUNTDOWN (1968) Astronauts prepare for a flight to the moon in this film that is more of a serious study than science-fiction. A big budget movie with a "big" director...Robert Altman! With stars you would never expect to see on *The World Beyond*, including Robert Duvall, James Caan, Joanna Moore, Ted Knight. Again, KPHO did much editing to squeeze this 101-minute film into a 90 minute time slot with commercials.

October 25—THE PHANTOM PLANET (1961) Rerun. Imagine *The World Beyond* following COUNTDOWN with this!

November 1—DRACULA (1931) Rerun.

November 8—HOUSE OF FRANKENSTEIN (1944) The Frankenstein monster, the Wolf Man, and Dracula all appear in this Universal monster mash which has been derided by many fans, but is still fun to watch. With Boris Karloff, Lon Chaney Jr., John Carradine, Glenn Strange, J. Carrol Naish, Lionel Atwill, George Zucco.

November 15—X THE UNKNOWN (1956) Rerun.

November 22—JOURNEY TO THE SEVENTH PLANET (1961) Rerun.

November 29—THE ANGRY RED PLANET (1959) Rerun.

December 6—DRACULA'S DAUGHTER (1936) Rerun.

December 13—SON OF DRACULA (1943) Rerun.

December 20—GHOSTS ON THE LOOSE (1943) An entry in the old Bowery Boys series, though at this time they were briefly being billed as The East Side Kids. Here, the tough guys discover that a reputedly haunted house is actually a front for Nazi spies. Only recommended to the Boys' decreasing number of fans. With Leo Gorcey, Huntz Hall, Bobby Jordan, Bela Lugosi, and a small bit by Ava Gardner before she hit the big time.

December 27—FRANKENSTEIN CONQUERS THE WORLD (1965) Rerun.

1976

January 3—THE GORGON (1964) Rerun.

January 10—CURSE OF THE MUMMY'S TOMB (1964) Rerun.

January 17—INVASION OF THE NEPTUNE MEN (1964) Rerun.

January 24—MONSTER ZERO (1968) Rerun.

January 31—VALLEY OF THE DRAGONS (1961) Rerun.

February 7—THE STRANGER (1973) An astronaut lands on another planet nearly identical to Earth, while agents of that world's government try to kill him. With Glenn Corbett, Cameron Mitchell, Sharon Acker, Dean Jagger, Lew Ayres, George Coulouris. Originally intended as a pilot to a proposed TV series that never made it, this film's airing on *The World Beyond* was significant— it was the first film made directly for television to show on the program, and it was also the first film from the 1970s to air on *World Beyond* as well. It could be said that *The World Beyond* entered the modern era with this film.

February 14—WILD, WILD PLANET (1965) Exceedingly dull Italian-made sci-fi about a sexy female alien who kidnaps Earth scientists by shrinking them. Nowhere near as fun as it sounds. With Tony Russell, Franco Nero.

February 21—GENESIS II (1973) A man wakes up from suspended animation and finds World War III has occurred, with the usual mutants running around causing mayhem. A made-for-tv movie produced by Gene Roddenberry as a pilot for a proposed series that never materialized. With Alex Cord, Mariette Hartley, Ted Cassidy.

February 28—I, MONSTER (1971) Retelling of the Dr. Jekyll and Mr. Hyde story, rather notorious in its day for keeping the names of the story's supporting characters, but changing the leads to Dr. Marlowe and Mr. Blake! If you can overlook that nonsense, this is not a bad rendition at all. With Christopher Lee, Peter Cushing, Mike Raven.

March 6—ATOMIC RULERS OF THE WORLD (1964) Rerun.

March 13—GHOST OF FRANKENSTEIN (1942) Fourth entry in Universal's Frankenstein series finds the Monster's friend Ygor (from SON OF FRANKENSTEIN, which never aired on *The World Beyond*) forcing the son of Dr. Frankenstein to transplant a new brain into the undying creature. Kind of fun, but most film historians believe the Universal monster series began its downward spiral with this one. With Sir Cedric Hardwicke, Evelyn Ankers, Lon Chaney Jr. (as the Monster), Bela Lugosi, Lionel Atwill, Ralph Bellamy.

March 20—PLANET OF BLOOD (1966) Rerun.

March 27—THE MUMMY'S HAND (1940) Archaeologists unearth a cursed tomb, and a shambling mummy starts terrorizing them. This Universal production is not really a sequel to THE MUMMY (1932), but was the beginning of the Kharis series—this film spawned three sequels which oddly never aired on *The World Beyond*. With Dick Foran, Peggy Moran, Wallace Ford, George Zucco, Eduardo Cianelli, Tom Tyler (as the mummy).

April 3—GODZILLA'S REVENGE (1967) Rerun.

April 10—IT'S ALIVE! (1968) Rerun.

April 17—MUTINY IN OUTER SPACE (1965) Rerun.

April 24—THE TWO FACES OF DR. JEKYLL (1960) Another version of the famed story, this one from Hammer Films, with a few twists. With Paul Massie, Dawn Addams, Christopher Lee (surprisingly in a supporting role), Francis De Wolff.

May 1—FROM THE EARTH TO THE MOON (1958) Rerun.

May 8—CURSE OF THE FLY (1965) Rerun.

May 15—VAMPIRE CIRCUS (1972) A 19th century traveling circus, run by vampires, travels from town to town in Eastern Europe! Despite being produced by Hammer Films, this one is not so easy to find today. With Adrienne Corri, Laurence Payne, Thorley Walters, John Moulder Brown, Lynne Frederick, David Prowse (the future Darth Vader!).

May 22—THE 4D MAN (1959) Rerun.

May 29—KUNG FU (1972) Made-for-tv movie was the pilot for the popular series with David Carradine as a Shao-Lin priest from China meandering across the 19th century American West, battling oppression and bigotry. Also with Keith Carradine in an unbilled bit, Barry Sullivan, Albert Salmi, Philip Ahn, Keye Luke, Richard Loo, Victor Sen Yung, Benson Fong. As this film is neither horror, science-fiction, or fantasy, it was a very strange choice to air on *The World Beyond*, and it is not known what KPHO programmers were thinking. In *The World Beyond* introduction, KPHO announcer Stu Tracy listed Keith Carradine before David, perhaps because Keith was enjoying some popularity as a singer at this time?

June 5—FRANKENSTEIN MEETS THE WOLF MAN (1943) Rerun.

June 12—THE MUMMY (1932) Rerun.

June 19—THE INVISIBLE MAN (1933) Rerun.

June 26—FRANKENSTEIN MEETS THE SPACE MONSTER (1965) Rerun.

July 3—THE HUMAN DUPLICATORS (1964) Rerun.

July 10—THE PEOPLE (1972) A made-for-tv movie in which a schoolteacher moves to a small town and discovers the unfriendly locals are actually aliens with ESP and other powers! With Kim Darby, William Shatner, Dan O'Herlihy, Diane Varsi.

July 17—THE INCREDIBLE SHRINKING MAN (1957) Rerun.

July 24—BEYOND THE TIME BARRIER (1959) Rerun.

July 31—THE BLOB (1958) Rerun.

August 7—SPOOKS RUN WILD (1941) Another entry in the Bowery Boys series, during the period when they were being billed as the East Side Kids. This time, they investigate a haunted house. With Leo Gorcey, Huntz Hall, Bela Lugosi, Angelo Rossito. This one and GHOSTS ON THE LOOSE were the only Bowery Boys films to ever air on *The World Beyond*, probably because Bela Lugosi appeared with them in both.

August 14—FIRST MAN INTO SPACE (1959) Rerun.

August 21—GHOSTS ON THE LOOSE (1943) Rerun.

August 28—PLANETS AGAINST US (1961) Rerun.

September 4—VOODOO MAN (1944) Mad doctor and his two assistants kidnap women and turn them into zombies. Produced by the extreme low-budget studio, Monogram Pictures, whose films are so terrible they are almost legendary. With Bela Lugosi, John Carradine, George Zucco.

September 11—FRANKENSTEIN CONQUERS THE WORLD (1966) Rerun.

September 18—GODZILLA VS. THE THING (1964) Rerun.

September 25—PLANET ON THE PROWL (1965) Rerun.

October 2—INVISIBLE INVADERS (1959) Space aliens take over the corpses of Earth's dead to conquer the world! With John Agar, John Carradine, Robert Hutton. Directed by Edward L. Cahn.

October 9—THE UNDERWATER CITY (1962) Rerun.

October 16—INVASION OF THE STAR CREATURES (1961) Rerun.

October 23—THE MAN FROM PLANET X (1951) Rerun.

October 30—CREATION OF THE HUMANOIDS (1962) Rerun.

November 6—MAN WITH THE SYNTHETIC BRAIN (1969) A mad scientist creates a zombie in this dreadful, low-budget opus that must be seen to be believed. Director Al Adamson (famous for his rock bottom films) patched together footage from two of his other films with some new footage he had shot in order to make this! It played theatres and drive-ins under a myriad of different names, including BLOOD OF GHASTLY HORROR. With John Carradine, Tommy Kirk, Kent Taylor, Regina Carrol, Rich Smedley.

November 13—WONDER WOMAN (1974) The famous heroine from DC Comics cracks an espionage ring in this made-for-tv movie. This was made several years before the popular tv series with Lynda Carter appeared. This one stars Cathy Lee Crosby in the lead (with a different costume than Wonder Woman is known for), Ricardo Montalban, Andrew Prine.

November 20—PLANET EARTH (1974) A 20th century man wakes up in the future to find that women have taken over the world and enslaved men! Naturally, he leads a rebellion. Made for tv film (clearly made in response to the feminist movement of the 1970s) was the pilot for a prospective tv series that never materialized, and has some of the same characters as producer Gene Roddenberry's GENESIS II which aired earlier in the year on *The World Beyond*. With John Saxon, Diana Muldaur, Janet Margolin, Ted Cassidy.

November 27—THE CREEPING UNKNOWN (1955) A scientist, Professor Quatermass, helps track down an astronaut who has turned into a monster after returning from space. First in a series of British films with the Professor as the lead character. With Brian Donlevy, Jack Warner, Lionel Jeffries.

December 4—THE GIANT BEHEMOTH (1959) Rerun.

December 11—QUEEN OF OUTER SPACE (1958) Rerun.

December 18—WAR OF THE SATELLITES (1958) Rerun.

December 25—THE MAGIC VOYAGE OF SINBAD (1962) A Yugoslavian-made fantasy, in which a warrior searches for the fabled bird of happiness to bring to his village. The lead character is Sinbad only courtesy of English-dubbing from the American distributor, and cast member names were similarly anglicized for release here. With Edward Stolar and Anna Larion—undoubtedly not their real names.

1977

January 1—WORLD WITHOUT END (1956) Rerun.

January 8—THE MAGNETIC MONSTER (1953) Rerun.

January 15—RED PLANET MARS (1952) Rerun.

January 22—X THE UNKNOWN (1956) Rerun.

January 29—JOURNEY TO THE SEVENTH PLANET (1961) Rerun.

February 5—KUNG FU (1972) Rerun of a film that wasn't suited to *The World Beyond* in the first place!

February 12—WAR-GODS OF THE DEEP (1965) Rerun.

February 19—THE INDESTRUCTIBLE MAN (1956) Mad scientist brings an executed murderer back to life, who goes on a killing spree. Considered by many to be "so bad it's entertaining," and fans of low-budget 1950s sci-fi flicks have a soft spot for this. With Lon Chaney Jr., Ross Elliott, Casey Adams, Robert Shayne.

February 26—GENESIS II (1973) Rerun.

March 5—THE WOLF MAN (1941) Rerun.

March 12—FRANKENSTEIN MEETS THE WOLF MAN (1943) Rerun.

March 19—THE INVISIBLE MAN (1933) Rerun.

March 26—THE MUMMY (1932) Rerun.

April 2—THE ANGRY RED PLANET (1959) Rerun.

April 9—THE BRAIN FROM PLANET AROUS (1957) Evil space alien who looks like a giant floating brain with eyes takes over the body of a nuclear scientist, while a good-guy brain possesses a dog while trying to nail the villain—and it is played fairly straight!!! Another film that must be seen to be believed. With John Agar, Joyce Meadows, Robert Fuller.

April 16—THE PHANTOM PLANET (1961) Rerun.

April 23—BEYOND THE TIME BARRIER (1960) Rerun.

April 30—THE STRANGER (1973) Rerun.

May 7—THE INCREDIBLE SHRINKING MAN (1957) Rerun.

May 14—GORGO (1961) Rerun.

May 21—CONQUEST OF THE PLANET OF THE APES (1972) The fourth entry in the popular series (the first three never aired on *The World Beyond*, though they appeared on some of KPHO's other movie shows). This depicts the rise of the ape ruler Caesar and shows how, in the future, apes came to rule the Earth. With Roddy McDowall, Don Murray, Ricardo Montalban, Severn Darden.

May 28—DINOSAURUS! (1960) Rerun.

June 4—DRACULA'S DAUGHTER (1936) Rerun.

June 11—GHOST OF FRANKENSTEIN (1942) Rerun.

June 18—THE MUMMY'S HAND (1940) Rerun.

June 25—NIGHT MONSTER (1942) Rerun.

July 2—INVASION OF THE STAR CREATURES (1961) Rerun.

July 9—CURSE OF THE MUMMY'S TOMB (1964) Rerun.

July 16—ATTACK OF THE CRAB MONSTERS (1957) People stranded on an Island are beset by giant, very intelligent crabs who assimilate the knowledge of their victims! Directed on a very low-budget by Roger Corman. With Richard Garland, Pamela Duncan, Russell Johnson, Ed Nelson, Jonathan Haze, Mel Welles.

July 23—CURSE OF THE FLY (1966) Rerun.

July 30—INVASION OF THE STAR CREATURES (1961) Rerun—twice in one month? It may have been a scheduling error in the tv listings, and something else may have been run one of the times. There is no way to know.

August 6—FACE OF FIRE (1959) A man is disfigured while rescuing a child from a fire, and the townspeople turn against him because of his looks. A straight, meaningful drama, probably mistaken for a horror film by KPHO programmers because of its title and its low budget. With Cameron Mitchell, James Whitmore.

August 13—MUTINY IN OUTER SPACE (1965) Rerun.

August 20—INVISIBLE INVADERS (1959) Rerun.

August 27—THE GIANT BEHEMOTH (1959) Rerun.

September 3—THE CREEPING UNKNOWN (1955) Rerun.

September 10—HORROR ISLAND (1941) Lesser known Universal horror film in which a motley group of people are stranded in a castle on an Island, and are menaced by a mysterious killer. With Dick Foran, Leo Carrillo, Peggy Moran, Fuzzy Knight.

September 17—FROM HELL IT CAME (1957) Research scientists on an Island are menaced by a giant, walking tree! The creature, designed by Paul Blaisdell, is one of the most absurd-looking monsters in the history of movies (no mean feat!). With Tod Andrews, Tina Carver.

September 24—DONOVAN'S BRAIN (1953) Rerun.

October 1—WAR OF THE SATELLITES (1958) Rerun.

October 8—SON OF DRACULA (1943) Rerun.

October 15—THE UNEARTHLY (1957) Mad scientist has a basement full of deformed mutants from his failed experiments, who eventually turn on him. Fans of rock-bottom horror films from the 1950s are rather fond of this one; I recall thinking it was quite dull. Judge for yourself. With John Carradine, Allison Hayes, Myron Healy, Tor Johnson.

October 22—THE HUMAN DUPLICATORS (1965) Rerun.

October 29—UNKNOWN WORLD (1951) Because of the threat of the Red Menace, scientists burrow into the Earth to find a hollow haven for mankind in case of a nuclear war. A very outdated relic of a bygone era, filmed in Carlsbad Caverns. With Bruce Kellogg, Marilyn Nash.

November 5—GODZILLA VS. THE SEA MONSTER (1967) The famed Japanese monster battles a creature named Ebirah, who looks like a giant lobster (he alone is worth watching the movie for!). Mothra also appears. Stock footage of the monsters fighting was re-used in GODZILLA'S REVENGE.

November 12—THE LOVE WAR (1970) Reportedly the first made-for-tv movie with space aliens (though they had appeared in various episodes of tv series), as two of them from rival planets decide to take human form and settle matters once and for all between them on neutral Earth. With Lloyd Bridges and Angie Dickinson!

November 19—MOON OF THE WOLF (1972) Made-for-tv movie with a Sheriff hunting a werewolf in the Bayou. With David Janssen, Bradford Dillman, Barbara Rush, John Beradino, Geoffrey Lewis.

November 26—BATTLE FOR THE PLANET OF THE APES (1973) Fifth and final entry in the famed original series, with ape leader Caesar battling a rebellion among the gorillas. Great cast, including Roddy McDowall, Claude Akins, Severn Darden, Paul Williams (!), Lew Ayres, John Huston.

December 3—THE INVISIBLE MAN RETURNS (1940) Rerun.

December 10—HOUSE ON HAUNTED HILL (1959) One of producer/director William Castle's best known films, with group of people agreeing to spend the night at a haunted house with the promise of a fortune if they make it through the night. With Vincent Price, Richard Long, Carol Ohmart, Elisha Cook Jr. I liked it well enough as a kid, but it looked far less impressive when I saw it again many years later.

December 17—THE 4D MAN (1959) Rerun.

December 24—THE MAGIC VOYAGE OF SINBAD (1962) Rerun.

December 31—EARTH VS. THE FLYING SAUCERS (1956) Rerun.

1978

January 7—IT! THE TERROR FROM BEYOND SPACE (1958) Bloodthirsty Martian monster stows away on a rocket ship and picks off the astronauts! The monster was designed by Paul Blaisdell, and played by Ray "Crash" Corrigan, who had been starring in serials 20 years earlier. Directed by Edward L. Cahn. Also with Marshall Thompson, Kim Spaulding, Dabbs Greer, Ann Doran, Paul Langton. Some sources say this film was the inspiration for the original ALIEN in 1979, but that's hard to believe.

January 14—THE MAN FROM PLANET X (1951) Rerun.

January 21—THE MAGNETIC MONSTER (1953) Rerun.

January 28—RIDERS TO THE STARS (1954) Rerun.

February 4—GHIDRAH, THE THREE-HEADED MONSTER (1965) Debut of the title character in this Japanese monster film, with good monsters Godzilla, Rodan, and Mothra out to stop his rampage. Ghidrah returned in many follow-up films from Toho Studios, including MONSTER ZERO which had already been shown on *The World Beyond*.

February 11—KUNG FU (1972) Rerun; third and final showing on *The World Beyond* of a film that was a very strange choice for the program to begin with.

February 18—SON OF GODZILLA (1968) UN scientists on an Island are beset by giant monsters, including the very cute title character who blows smoke rings instead of his dad's radioactive breath. Silly but kind of fun.

February 25—THE INCREDIBLE SHRINKING MAN (1957) Rerun.

March 4—THE BRAIN FROM PLANET AROUS (1957) Rerun.

March 11—THE PHANTOM PLANET (1961) Rerun.

March 18—BEYOND THE TIME BARRIER (1960) Rerun.

March 25—X THE UNKNOWN (1956) Rerun.

April 1—THE PHARAOH'S CURSE (1957) An Egyptian expedition is beset by supernatural creatures. Extremely low-budget and dull. With Mark Dana, Diane Brewster, Ziva Rodann.

April 8—CURSE OF THE MUMMY'S TOMB (1964) Rerun.

April 15—CURSE OF THE FACELESS MAN (1958) An expedition at Pompeii is menaced by a resurrected ancient citizen, encased in lava. Directed by Edward L. Cahn. With Richard Anderson, Elaine Edwards, Adele Mara.

April 22—VOODOO ISLAND (1957) One of Boris Karloff's worst movies, as he plays a skeptic who travels to an Island to debunk claims of voodoo practice and lives to regret it. Very low-budget outing, also with Rhodes Reason, Elisha Cook Jr., Murvyn Vye.

April 29—WILD, WILD PLANET (1965) Rerun.

May 6—GODZILLA VS. THE SEA MONSTER (1967) Rerun.

May 13—GENESIS II (1973) Rerun.

May 20—CONQUEST OF THE PLANET OF THE APES (1973) Rerun.

May 27—THE GORGON (1964) Rerun.

June 3—INVASION OF THE STAR CREATURES (1961) Rerun.

June 10—ATTACK OF THE CRAB MONSTERS (1957) Rerun.

June 17—THE NEANDERTHAL MAN (1953) Rerun.

June 24—MARK OF THE VAMPIRE (1957) Rerun of a film that had aired on *The World Beyond* in 1965 under its alternate title, THE VAMPIRE.

July 1—THE STRANGER (1973) Rerun.

July 8—THE MYSTERIANS (1957) Rerun.

July 15—RED PLANET MARS (1952) Rerun.

July 22—WORLD WITHOUT END (1956) Rerun.

July 29—THE BEGINNING OF THE END (1957) Thanks to atomic radiation, giant grasshoppers invade Illinois! Produced and directed by Bert I. Gordon. With Peter Graves, Peggie Castle, Morris Ankrum.

August 5—NIGHT KEY (1937) Boris Karloff plays a kindly inventor who goes after his partner for stealing his new burglar alarm invention, and winds up forced to help crooks rob businesses. Strictly a low-grade crime drama; this was probably scheduled on *The World Beyond* due to Karloff's name—perhaps KPHO programmers mistook it for a horror movie for that reason. Also with Warren Hull, Hobart Cavanaugh.

August 12—I BURY THE LIVING (1958) Cemetery manager believes pushing pins into reserved plots on the cemetery's map causes people to die. This small film has developed a mild cult following over the years. With Richard Boone, Theodore Bikel.

August 19—THE MUMMY'S HAND (1940) Rerun. This was oddly the last of Universal's famous monster films to be shown on *The World Beyond*.

August 26—THE BEAST OF HOLLOW MOUNTAIN (1956) Dinosaur menaces people in the Mexican wilderness—more or less, as he doesn't actually appear until the last half hour, making for a pretty slow-moving film. With Guy Madison, Patricia Medina.

September 2—MUTINY IN OUTER SPACE (1965) Rerun.

September 9—THE CURSE OF DRACULA (1958) The Count puts the bite on people in then-contemporary California in this low-budget opus. Also released under the name RETURN OF DRACULA. With Francis Lederer, Ray Stricklyn.

September 16—THE UNDERWATER CITY (1962) Rerun.

September 23—THE BLOB (1958) Rerun.

September 30—VALLEY OF THE DRAGONS (1961) Rerun.

October 7—CASTLE OF TERROR (1963) Rerun of a film that had aired on *The World Beyond* in 1970 under its alternate title, CASTLE OF BLOOD.

October 14—FACE OF FIRE (1959) Rerun.

October 21—MACABRE (1958) After his family is murdered, a small town doctor tries to discover if his daughter has been buried alive. Notable as producer/director William Castle's first horror movie. With William Prince, Jim Backus.

October 28—FROM HELL IT CAME (1957) Rerun.

November 4—GODZILLA, KING OF THE MONSTERS (1956) Rerun.

November 11—GODZILLA VS. THE SEA MONSTER (1967) Rerun.

November 18—GHIDRAH, THE THREE HEADED MONSTER (1965) Rerun.

November 25—TERROR OF MECHAGODZILLA (1978) Godzilla battles a robot duplicate of himself. Notable as the only film to air on *The World Beyond* the same year it was made; upon completion in Japan, it was immediately sold to television in America while simultaneously receiving a very limited American theatrical release under the alternate title TERROR OF GODZILLA. It is a direct sequel to GODZILLA VS. THE COSMIC MONSTER, which would air on *The World Beyond* in 1980. This was also the final film in Toho's original Godzilla series; they would start up a new round years later.

December 2—THE WEREWOLF (1956) A werewolf terrorizes a small town. With Don Megowan, Joyce Holden, Steven Ritch.

December 9—THE INVISIBLE INVADERS (1959) Rerun.

December 16—*The World Beyond* was pre-empted by the Garden State Bowl (a ball game) and a rerun of an Adam-12 episde.

December 23—THE THREE STOOGES IN ORBIT (1962) Rerun.

December 30—THE CREEPING UNKNOWN (1955) Rerun.

1979

January 6—DONOVAN'S BRAIN (1953) Rerun.

January 13—IT! THE TERROR FROM BEYOND SPACE (1958) Rerun.

January 20—THE BLACK SLEEP (1956) Another mad scientist and his experiments; he keeps his failed mutant subjects chained in the basement. A large cast of "horror stars" make this opus a lot of fun. With Basil Rathbone, Bela Lugosi, Lon Chaney Jr., John Carradine, Akim Tamiroff, Tor Johnson.

January 27—RIDERS TO THE STARS (1954) Rerun.

February 3—RODAN (1956) Rerun.

February 10—GODZILLA, KING OF THE MONSTERS (1956) Rerun.

February 17—WONDER WOMAN (1974) Rerun.

February 24—GODZILLA'S REVENGE (1967) Rerun.

March 3—WAR OF THE SATELLITES (1958) Rerun.

March 10—THE MAGNETIC MONSTER (1953) Rerun.

March 17—BRIDE OF THE GORILLA (1951) Who could resist a title like that? A plantation owner starts turning into a killer ape after being cursed by a witch doctor. Directed by Curt Siodmak.

With Lon Chaney Jr., Barbara Payton, Raymond Burr, Tom Conway.

March 24—EARTH VS. THE FLYING SAUCERS (1956) Rerun.

March 31—THE MAN FROM PLANET X (1951) Rerun.

April 7—BEYOND THE TIME BARRIER (1960) Rerun.

April 14—THE MAGIC VOYAGE OF SINBAD (1962) Rerun.

April 21—UNKNOWN WORLD (1951) Rerun.

April 28—THE PHANTOM PLANET (1961) Rerun.

May 5—MONSTER ZERO (1968) Rerun.

May 12—PHASE IV (1974) Swarms of ants band together and start attacking humans in an effort to take over the world. Unusually grisly for this kind of sci-fi. With Nigel Davenport, Lynne Frederick, Michael Murphy.

May 19—BATTLE OF THE WORLDS (1961) Low-budget Italian sci-fi film with a planet run by computers threatening to destroy Earth. Notable for starring Claude Rains, in what was a new career low for him. Also with Bill Carter, Maya Brent, Umberto Orsini.

May 26—WAR OF THE GARGANTUAS (1967) Rerun.

June 2—ATTACK OF THE CRAB MONSTERS (1957) Rerun.

June 9—THE BRAIN FROM PLANET AROUS (1957) Rerun.

June 16—*The World Beyond* was pre-empted by the Portland Rose Festival Parade.

June 23—MARK OF THE VAMPIRE (1957) Rerun.

June 30—THE ANGRY RED PLANET (1959) Rerun.

July 7—DON'T BE AFRAID OF THE DARK (1973) A young couple moves to an old house, and discover that evil gnome-like creatures live under it! A made-for-tv movie directed by John Newland, which incredibly was remade as a big-budget theatrical release in 2011! With Kim Darby, Jim Hutton, William Demarest.

July 14—FIRST SPACESHIP ON VENUS (1962) Rerun.

July 21—VARAN THE UNBELIEVABLE (1962) Rerun.

July 28—THE NAVY VS.THE NIGHT MONSTERS (1966) Rerun.

August 4—WOMEN OF THE PREHISTORIC PLANET (1965) Rerun.

August 11—MAN WITH THE SYNTHETIC BRAIN (1969) Rerun. I have a personal memory of watching the rerun on *The World Beyond* and realizing that KPHO had gotten the film reels mixed up and were running it out of sequence. It hardly mattered since the film is so incoherent anyway!

August 18—THE GIANT BEHEMOTH (1959) Rerun.

August 25—BEGINNING OF THE END (1957) Rerun.

September 1—THE GORGON (1964) Rerun.

September 8—X THE UNKNOWN (1956) Rerun.

September 15—RED PLANET MARS (1952) Rerun.

September 22—THE NEANDERTHAL MAN (1953) Rerun.

September 29—THE BEAST OF HOLLOW MOUNTAIN (1956) Rerun.

October 6—THE NAVY VS. THE NIGHT MONSTERS (1966) Rerun.

October 13—BEYOND THE TIME BARRIER (1960) Rerun.

October 20—PHASE IV (1974) Rerun.

October 27—GENESIS II (1973) Rerun.

November 3—GHIDRAH, THE THREE HEADED MONSTER (1965) Rerun.

November 10—GODZILLA VS. THE SEA MONSTER (1967) Rerun.

November 17—SON OF GODZILLA (1968) Rerun.

November 24—GODZILLA VS. MEGALON (1973) Godzilla and a giant robot named Jet Jaguar battle the monsters Megalon and

Gigan. Fans of the Japanese fighting monsters consider this to be one of the worst in the series.

December 1—WORLD WITHOUT END (1956) Rerun.

December 8—THE CREEPING TERROR (1964) Ultra-low-budget film about an alien fungus monster (who looks like a giant carpet) that invades the Earth. The entire movie is narrated with no character dialogue, reportedly because the filmmakers accidentally destroyed the audio after filming! This regularly appears on lists of the worst movies of all time, and many would consider this to be the worst film to ever air on *The World Beyond* (no mean feat!). With Vic Savage, Shannon O'Neill.

December 15—PLANET EARTH (1974) Rerun.

December 22—CONQUEST OF THE PLANET OF THE APES (1972) Rerun.

December 29—WAR OF THE SATELLITES (1958) Rerun.

1980

January 5—INVASION OF THE STAR CREATURES (1961) Rerun.

January 12—THE CREEPING UNKNOWN (1955) Rerun.

January 19—CURSE OF DRACULA (1958) Rerun.

January 26—JOURNEY TO THE SEVENTH PLANET (1961) Rerun.

February 2—GORGO (1980) Rerun.

February 9—THE LOVE WAR (1970) Rerun.

February 16—BATTLE OF THE WORLDS (1961) Rerun.

February 23—GODZILLA, KING OF THE MONSTERS (1956) Rerun.

March 1—QUEEN OF OUTER SPACE (1958) Rerun.

March 8—THE INCREDIBLE SHRINKING MAN (1957) Rerun.

March 15—INVISIBLE INVADERS (1959) Rerun.

March 22—FROM HELL IT CAME (1957) Rerun.

March 29—BATTLE BENEATH THE EARTH (1968) The Red Chinese build a series of tunnels with which to invade America, while a small U.S. Army detachment tries to stop them! Needless to say, with the U.S. Government's more favorable treatment of China at the present time, this is a film that would not be made today. With Kerwin Mathews, Viviane Ventura.

April 5—THE DAY THE EARTH FROZE (1959) Rerun.

April 12—LEGEND OF BOGGY CREEK (1976) A Bigfoot-type critter stalks the backwoods of Arkansas, terrifying the locals! Very low-budget film, produced and directed by Charles B. Pierce, was an unexpected success and spawned two sequels! Some sources say it was made as early as 1972. With Willie Smith, John Nixon. In his *World Beyond* introduction, Stu Tracy somehow erred by stating that this starred David Hess and Lucy Grantham—they were actually the stars of Wes Craven's infamous LAST HOUSE ON THE LEFT, but not in this Boggy Creek film!

April 19—GHOSTS THAT STILL WALK (1977) Family consults a psychic after their son is possessed by a supernatural force. This low-grade horror film, which includes a scene of rocks attacking the family's house (!), was fraudulently sold as a documentary during the period in the 1970s when the market was bombarded by documentaries about the paranormal. In fact, this seems to have been part of a syndicated package of such documentaries that KPHO acquired at this time and were running on their various movie programs. With Ann Nelson.

April 26—THE LOST CITY OF ATLANTIS (1977) A paranormal documentary on the obvious subject, made during the period when documentaries about the supernatural were popular. This was the only true documentary to ever air on *The World Beyond*.

May 3—SSSSSSS (1973) A mad scientist experiments with turning human beings into King Cobras! A rather repellent approach to its plot makes this one far less hokey than it sounds—in fact, an argument could be made this film was more "adult" than most of the films booked on *The World Beyond* up to this time. With Strother Martin, Dirk Benedict, Heather Menzies.

May 10—GODZILLA VS. THE COSMIC MONSTER (1974) Godzilla battles Mechagodzilla, a robot duplicate of himself controlled by aliens. Meanwhile, a monster who looks like a giant shaggy dog (who seems to be named King Seesaw in the film; Wikipedia reports that Toho Studios later trademarked him in English as King Caesar) also appears, as does the monster Angurus. When this was syndicated to American television, it was accompanied by inexplicable erroneous information that it starred Jack Palance and Carol Lynley! This is not true, but oh, don't you wish it were? Unknowing, Stu Tracy repeated it in his introduction, and it was used in KPHO's ads for this film. This wrong information still appears on some lesser movie websites to this day.

Godzilla vs. the
Cosmic Monster

Starring Jack Palance
& Carol Lynley,
believe it or not.
10:30 Today

tv5
K P H O · P H O E N I X

Don't believe it. The "stars" were not actually in the film, but don't you wish they were? This was a piece of misinformation that made the rounds at the time, and it can still be found on some obscure movie websites.

May 17—COMEDY OF TERRORS (1964) A dark comedy about a corrupt funeral home is given some energy by a large cast of well-known horror stars, and is quite entertaining. With Vincent Price, Peter Lorre, Boris Karloff, Basil Rathbone, Joe E. Brown. Screenplay by Richard Matheson, and directed by Jacques Tourneur.

May 24—KING KONG VS. GODZILLA (1963) The two monsters duke it out in one of the better known Japanese monster movies, with added footage of American actors shot for American release. With Michael Keith, Harry Holcombe, James Yagi.

May 31—THE DAY THE EARTH MOVED (1974) Two earthquake forecasters desperately try to persuade the skeptical residents of a small town to evacuate before "the big one" hits. Pretty much a straight drama, probably booked on *The World Beyond* by KPHO programmers who were misled by the title of this made-for-tv movie. With Jackie Cooper, Stella Stevens, Cleavon Little, William Windom, Beverly Garland, E.J. Andre, Sid Melton, Don Steele.

June 7—THE ANGRY RED PLANET (1959) Rerun.

June 14—BEYOND THE TIME BARRIER (1960) Rerun.

June 21—THE PHANTOM PLANET (1961) Rerun.

June 28—THE MAGNETIC MONSTER (1953) Rerun.

July 5—CREATURE WITH THE BLUE HAND (1967) German film based on an Edgar Wallace story, about a series of gruesome murders. Klaus Kinski plays identical twins, one of whom may be the killer! Also with Diana Kerner.

July 12—SON OF BLOB (1972) Long-belated sequel to THE BLOB (1958), and it is absolutely dreadful, despite a terrific cast of stars. Some of it seems to be intentionally played for laughs, but the whole film is so bad it is hard to tell for sure. Directed by tv actor Larry Hagman (who would later become J.R. Ewing on the DALLAS tv series), and he appears on screen as well. But how did he persuade Robert Walker, Godfrey Cambridge, Carol Lynley, Marlene Clark, Cindy Williams, Richard Stahl, Shelley Berman, Burgess Meredith (who had his name removed from the credits), Gerrit Graham, and Dick Van Patten to appear in this glop? Also released under the title, BEWARE! THE BLOB.

July 19—DRACULA VS. FRANKENSTEIN (1971) Very low-budget and incompetently made trash (but almost fascinating on that level) about Count Dracula forcing an aged Dr. Frankenstein to revive his monster. Directed by Al Adamson, the man who also gave us MAN WITH THE SYNTHETIC BRAIN that was shown on *The World Beyond*. This was sadly the final movie for both J. Carrol Naish and Lon Chaney Jr., who both look terrible, ravaged by illness and age. Also with Russ Tamblyn, Jim Davis, Angelo Rossito, Regina Carrol, Zandor Vorkov, John Bloom, Anthony Eisley, Forrest J. Ackerman.

July 26—GODZILLA VS. THE SMOG MONSTER (1972) At the peak of the 1970s ecology movement, Toho Studios in Japan pitted Godzilla against Hedorah, a monster spawned by toxic waste and pollution! Godzilla's fans have a low opinion of this one, but I have a soft spot for it, as the first Godzilla movie I ever saw—at a kiddie matinee when it was fairly new, at which Wallace and Ladmo appeared.

August 2—KONGA (1961) A mad scientist turns a chimpanzee into a giant ape that goes berserk in London. With Michael Gough, Margo Johns.

August 9—THE INDESTRUCTIBLE MAN (1956) Rerun.

August 16—ATTACK OF THE MONSTERS (1968) Japanese movie in which two children are kidnapped in a UFO by space alien women who eat brains. The friendly giant turtle Gamera comes to their rescue, and battles Guiron, a beast with a head shaped like a knife. With Christopher Murphy. This was the first film in the popular Gamera series to be shown on *The World Beyond*; he had been created by Toei Studios as competition for Godzilla movies. A change in Copyright holders in recent years resulted in this being retitled GAMERA VS. GUIRON.

August 23—THE MAGIC SERPENT (1966) A young warrior in Ancient times battles supernatural critters while trying to avenge his father's murder. A Japanese film that has some admirers.

August 30—GENESIS II (1973) Rerun.

September 6—THE BEAST OF HOLLOW MOUNTAIN (1956) Rerun.

September 13—THE GIANT BEHEMOTH (1959) Rerun.

September 20—MAN WITH THE SYNTHETIC BRAIN (1969) Rerun.

September 27—DONOVAN'S BRAIN (1953) Rerun.

October 4—THE BRAIN FROM PLANET AROUS (1957) Rerun. Notice this represented three 'brain' movies in a row! That had to be intentional.

October 11—RIDERS TO THE STARS (1954) Rerun.

October 18—THE NEANDERTHAL MAN (1953) Rerun.

October 25—GARGOYLES (1972) Made-for-tv film in which an anthropologist and his daughter are pursued by ancient, gargoyle monsters trying to retrieve the skull of one of them which he has discovered. With Cornel Wilde, Jennifer Salt, Bernie Casey, Grayson Hall.

November 1—INVADERS FROM MARS (1953) Only a little boy realizes that space aliens have taken over his town, possessing everyone except him! Regarded by many as an all-time sci-fi classic, and this even led to a big-budget remake in 1986! Myself, I found it wildly overrated and not much different from many low-grade yarns from the 1950s, so judge for yourself. Directed by William Cameron Menzies. With Arthur Franz, Helena Carter, Leif Erickson, Morris Ankrum, Milburn Stone.

November 8—GODZILLA ON MONSTER ISLAND (1972) Godzilla and the monster Angurus battle Ghidrah and a giant robotic bird named Gigan. Silly stuff; this is the one notorious for a few scenes of the monsters actually speaking to each other! Also known as GODZILLA VS. GIGAN.

November 15—THE ALIEN FACTOR (1977) Three alien monsters invade a small town in this film with almost no budget at all; it was shot in 16mm film and was almost a home movie that somehow got released. With Don Leifert, George Stover.

November 22—STAR PILOT (1967) Incomprehensible Italian sci-fi mess with aliens and UFOs flying around. This went unreleased in America until 1977, when it suddenly starting appearing in lesser theatres with a misleading ad campaign designed to cash in on the STAR WARS craze that had just erupted. With Kirk Morris, Gordon Mitchell.

November 29—FRANKENSTEIN CONQUERS THE WORLD (1966) Rerun.

December 6—UNKNOWN WORLD (1951) Rerun.

December 13—EARTH VS. THE FLYING SAUCERS (1956) Rerun.

December 20—BEGINNING OF THE END (1957) Rerun.

December 27—VARAN THE UNBELIEVABLE (1962) Rerun.

1981

January 3—WORLD WITHOUT END (1956) Rerun.

January 10—THE BLACK SLEEP (1956) Rerun.

January 17—THE MAN FROM PLANET X (1951) Rerun.

January 24—RED PLANET MARS (1952) Rerun.

January 31—PLANET EARTH (1974) Rerun.

February 7—THE THING (1951) Rerun.

February 14—GODZILLA VS. MEGALON (1973) Rerun.

February 21—DARK STAR (1974) Astronauts on a space-ship follow orders to bomb unstable planets, while dealing with space critters and a snarky computer with a mind of its own. Basically a spoof of the famed 2001:A SPACE ODYSSEY, and a surprisingly enjoyable one. This low-budget fare marked the first movie to be directed by John Carpenter. It has a great theme song that plays over the credits, called "Benson, Arizona"! With Brian Narelle, Dan O'Bannon (who went on to become a sci-fi author and screenwriter).

One of the biggest films to ever air on The World Beyond.

February 28—WESTWORLD (1973) In an amusement park in the future, human-like robots cater to the needs of rich tourists....until they start to malfunction. A major studio production, written and directed by Michael Crichton, and one of the "biggest" movies to ever air on *The World Beyond*. With Yul Brynner, Richard Benjamin, James Brolin. In 2016, Westworld was remade and rebooted as an HBO tv series!

March 7—GORGO (1961) Rerun.

March 14—THE UNEARTHLY (1957) Rerun.

March 21—MARK OF THE VAMPIRE (1957) Rerun.

March 28—FIRST SPACESHIP ON VENUS (1962) Rerun.

April 4—I BURY THE LIVING (1958) Rerun.

April 11—THE CREEPING UNKNOWN (1955) Rerun.

April 18—IT! THE TERROR FROM BEYOND SPACE (1958) Rerun.

April 25—THE WEREWOLF (1956) Rerun.

May 2—FROGS (1972) All kinds of swamp animals—frogs, snakes, lizards, etc.—attack the plantation of a local wealthy family. With Ray Milland, Sam Elliott, Joan Van Ark, Adam Roarke.

May 9—HORROR AT 37,000 FEET (1973) A made-for-tv movie in which a large cursed stone in the baggage compartment unleashes a demon on a 747 jet! With William Shatner, Buddy Ebsen, Chuck Connors, Roy Thinnes, France Nuyen, Tammy Grimes, Paul Winfield, Russell Johnson, H.M. Wynant, Robert Donner.

Frogs

They are trying to croak Sam Elliott & Joan Van Ark.

10:30 Today

tv5
KPHO · PHOENIX

May 16—INVADERS FROM MARS (1953) Rerun.

May 23—EMPIRE OF THE ANTS (1977) Giant monster ants snack on people after eating radioactive waste! Produced, written, and directed by Bert I. Gordon. With Joan Collins, Robert Lansing, Albert Salmi, John David Carson.

May 30—NIGHT SLAVES (1970) Space aliens hypnotize the residents of a small town to repair their spaceship in this made-for-tv movie. With James Franciscus, Lee Grant, Leslie Nielsen, Andrew Prine, Tisha Sterling.

June 6—HORROR HOUSE (1969) A mad killer picks off people at a party

EMPIRE OF THE ANTS

Robert Lansing, Joan Collins

10:30 Today

tv5
KPHO · PHOENIX

held in a reputedly haunted house. With Frankie Avalon (!), Jill Haworth, Dennis Price.

June 13—RETURN OF GIANT MAJIN (1966) A Japanese movie with a giant stone statue coming to life in Ancient times to help his impoverished people; sort of a variation of the Golem legend. This was the second of three Majin movies. The other two were never shown on *The World Beyond.*

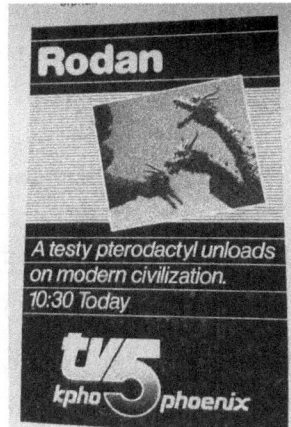

Rodan

A testy pterodactyl unloads on modern civilization.
10:30 Today

tv5
kpho phoenix

The photo in this ad is from the wrong movie, as it shows Rodan battling Ghidrah, who did not appear in the original 1956 RODAN film.

June 20—VILLAGE OF THE DAMNED (1960) A group of strange children with supernatural powers, all born at the same time, take over a town. Quite an excellent film, directed by Wolf Rilla. With George Sanders, Barbara Shelley.

June 27—RODAN (1956) Rerun.

July 4—TIME TRAVELERS (1976) Two men travel back in time to 1871 Chicago to try to prevent the Great Fire. Made-for-tv twaddle, produced by Irwin Allen as the pilot for a proposed tv series that never materialized. With Sam Groom, Tom Hallick, Richard Basehart, Francine York.

July 11—SON OF KONG (1933) Rerun.

July 18—STARSHIP INVASIONS (1977) A scientist is kidnapped by space aliens to help fight a war with evil space aliens. They are all played by actors in garish suits reminiscent of old 1950s sci-fi! Basically a STAR WARS rip-off wasting some excellent actors. With Robert Vaughn, Christopher Lee, Daniel Pilon, Helen Shaver.

July 25—FOOD OF THE GODS (1976) Rats, wasps, and various farm animals grow to giant size after eating a strange substance. Produced and directed by Bert I. Gordon. With Marjoe Gortner, Pamela Franklin, Ida Lupino, Ralph Meeker.

August 1—DESTROY ALL MONSTERS (1968) Female space aliens attempt to take over the Earth by controlling our planet's monsters. Virtually ALL of the Toho Studio's monsters appear in this film—Godzilla, Rodan, Mothra, Ghidrah, you name them. Even Godzilla's son who blows smoke rings appears! Their fans are rather fond of this one, even if it is silly.

August 8—INVISIBLE INVADERS (1959) Rerun.

August 15—CURSE OF DRACULA (1958) Rerun.

August 22—MONSTER ZERO (1968) Rerun.

August 29—THE MAGNETIC MONSTER (1953) Rerun.

September 5—QUEEN OF OUTER SPACE (1958) Rerun.

September 12—DONOVAN'S BRAIN (1953) Rerun.

September 19—CONQUEST OF THE PLANET OF THE APES (1972) Rerun.

September 26—THE BRAIN FROM PLANET AROUS (1957) Rerun.

October 3—CURSE OF THE SWAMP CREATURE (1966) Rerun.

October 10—YONGARY, MONSTER FROM THE DEEP (1968) A dinosaur goes on a rampage after being awakened by an Earthquake. This is a South Korean film, believe it or not, although it is virtually indistinguishable from most Japanese monster movies. With Oh Yung II, Moon Kang.

October 17—EYES BEHIND THE STARS (1972) A man goes on the run from Government assassins after discovering the plot to cover up the existence of UFOs. Low-budget, fairly obscure Italian film, with Martin Balsam, Robert Hoffman.

October 24—FANTASTIC INVASION OF PLANET EARTH (1966) Young couple is stranded in a deserted town, enclosed by a giant force field all around which they can't escape from. Alternate title, THE BUBBLE, was more accurate about the film's content. Written and directed by Arch Oboler. With Michael Cole, Deborah Walley.

October 31—THE PEOPLE THAT TIME FORGOT (1977) Search party visits an Island with prehistoric monsters while looking for an expedition that vanished years earlier. Direct sequel to THE LAND THAT TIME FORGOT (1975) which would air in 1982 on *The World Beyond*. With Patrick Wayne, Doug McClure, Sarah Douglas, Thorley Walters.

November 7—GODZILLA'S REVENGE (1967) Rerun.

November 14—SQUIRM (1976) Quite good but very grisly film about worms in the Deep South attacking humans. Some very effective shock scenes, not really suited for *The World Beyond*'s juvenile audience. With Don Scardino, Patricia Pearcy.

November 21—THE PIT AND THE PENDULUM (1961) One of director Roger Corman's loose adaptations of Edgar Allan Poe's stories. Descendent of an official of the Spanish Inquisition goes mad and starts torturing visitors in grisly ways. With Vincent Price, John Kerr, Barbara Steele, Luana Anders. Another film marking the perhaps unintentional shift to more adult films on *The World Beyond*.

November 28—THE FABULOUS WORLD OF JULES VERNE (1958) A Czech import, partly animated to make the action look like 19th century engravings, pertaining to several of Verne's stories. Released in America with English dubbing and phony Americanized cast names. This film has quite a few admirers. Directed by Karel Zeman, who made other similar films as well.

December 5—FRANKENSTEIN'S BLOODY TERROR (1968) The werewolf Waldemar Daninsky tries to find a cure for his lycanthropy, but the clinic where he goes is operated by vampires. With Paul Naschy, who was a big horror star in Spain (where this was made) but he did not catch on in other countries. He made several films as Daninsky the werewolf. The American release title was a complete fraud, as the film has absolutely nothing to do with Frankenstein. This was the only Naschy film to ever air on *The World Beyond*.

December 12—WAR-GODS OF THE DEEP (1966) Rerun.

December 19—GOOD AGAINST EVIL (1977) An author tries to rescue a woman from a Satanic cult that plans to have her bear the Devil's child in this made-for-tv movie. With Dack Rambo, Dan O'Herlihy, Richard Lynch, Kim Cattrall, Elyssa Davalos.

December 26—SON OF GODZILLA (1968) Rerun.

1982

January 2—THE CRIMSON CULT (1968) A witch who was burned at the stake 300 years earlier returns to take revenge on the descendants of her persecutors. A great cast in a remarkably incoherent film. With Boris Karloff, Christopher Lee, Barbara Steele, Michael Gough. When this British film was released in America after Karloff's death, it was advertised as his final movie. It wasn't; the great horror star had filmed scenes for four low-grade Mexican quickies after this.

January 9—REPTILICUS (1962) Rerun.

January 16—X-THE MAN WITH THE X-RAY EYES (1963) Rerun.

January 23—I, MONSTER (1971) Rerun.

January 30—THE EYE CREATURES (1965) The title creatures (played by men in ridiculous suits) invade the Earth from outer space. Directed by Larry Buchanan on very little budget. With John Ashley, Cynthia Hull.

February 6—THE BEAST MUST DIE (1974) Motley group of people staying on a wealthy benefactor's Island are told one of them is a werewolf, but which one? Sort of a werewolf version of an Agatha Christie mystery. With Calvin Lockhart, Peter Cushing, Charles Gray, Marlene Clark, Anton Diffring.

February 13—THE CONQUEROR WORM (1968) The witchfinder Matthew Hopkins puts suspected witches to death in Medieval

England. Excellent film, but very violent and gory; it was banned in several countries around the world. One of the most adult films to ever air on *The World Beyond*. With Vincent Price, Ian Ogilvy, Rupert Davies, Robert Russell. In Britain, it was released under the name WITCHFINDER GENERAL. For American distribution, it was retitled with the name of an Edgar Allan Poe poem to make potential patrons think it was part of Price's popular Poe adaptations.

February 20—THE LUCIFER COMPLEX (1978) An incoherent mess, consisting of footage from a different, uncompleted movie, and new footage shot to bring this to feature length—something about the cloning experiments of modern-day Nazis. With Robert Vaughn, Keenan Wynn, Aldo Ray, Leo Gordon.

February 27—GODZILLA, KING OF THE MONSTERS (1956) Rerun.

March 6—DEAD OF NIGHT (1977) Three short horror stories in this made-for-tv movie produced and directed by Dan Curtis and written by Richard Matheson. It was a pilot for a proposed series that never got off the ground. With Horst Buccholz, Ed Begley Jr., Anjanette Comer, Joan Hackett, Lee Harcourt Montgomery, Patrick Macnee, Ann Doran, E.J. Andre, Elisha Cook Jr.

GODZILLA
Raymond Burr
10:30 Today

tv5
KPHO PHOENIX

March 13—GAMERA VS. MONSTER X (1970) Japanese movie with the evil monster Giger laying eggs in Gamera's body so the babies can drink his blood and destroy him (yes, you read that correctly). Two little boys help the giant turtle save the day. A change in American distributors and copyright holders years later resulted in this being retitled GAMERA VS. GIGER.

March 20—THE PREMATURE BURIAL (1962) One of director Roger Corman's loose Poe adaptations, this one about a man who suffers from Taphophobia (a fear of being buried alive) being victimized by his greedy family. With Ray Milland, Hazel Court, Richard Ney, Heather Angel.

March 27—THE INCREDIBLE TWO-HEADED TRANSPLANT (1971) Wildly over-the-top horror film with a very interesting cast. Mad scientist grafts the head of an insane murderer onto the body of a retarded man! With Bruce Dern, Pat Priest, Casey Kasem (!), John Bloom.

April 3—YOG—MONSTER FROM SPACE (1971) The title character is a giant octopus from outer space, who goes berserk along with other critters. Yog is by far the most absurd-looking monster ever created by Toho Studios in Japan, and that makes this film very hard to dislike.

April 10—LEGEND OF BOGGY CREEK (1976) Rerun.

April 17—WHERE TIME BEGAN (1978) A well-intentioned Spanish film adaptation of Jules Verne's "Journey To The Center Of The Earth", defeated by its low-budget. Directed by Juan Piquer Simon. With Kenneth More, Jack Taylor, Pep Munne.

April 24—ATTACK OF THE MONSTERS (1968) Rerun. This aired in place of CRAZE, a 1974 Jack Palance film about an insane murderer that had been originally scheduled. It is unknown why KPHO pulled it; perhaps someone alerted them that it was too adult for *The World Beyond*. If so, it was hardly more so than THE CONQUEROR WORM or other films that were soon to be scheduled. CRAZE later aired late at night on KPHO's *Movietime* show.

May 1—THE LAND THAT TIME FORGOT (1975) An expedition gets stranded on an Island inhabited by prehistoric dinosaurs. The sequel, THE PEOPLE THAT TIME FORGOT, had already been shown on *The World Beyond* in 1981. With Doug McClure, Susan Penhaigon, John McEnery.

May 8—TENTACLES (1977) A giant octopus picks off the local beach population! Bizarre casting almost makes this worth seeing. With John Huston, Shelley Winters, Henry Fonda, Bo Hopkins, Cesare Danova. Fonda, in an overblown cameo, was the biggest star in the film (and perhaps the biggest star to ever be seen on *The World Beyond!*), but interestingly, Stu Tracy's introduction to the film failed to mention him.

May 15—THE CAT CREATURE (1973) Quite good made-for-tv chiller about mysterious deaths that follow the theft of some cursed Egyptian artifacts. Written by Robert Bloch and directed by Curtis Harrington. With Meredith Baxter, Stuart Whitman, Gale Sondergaard, John Carradine, Keye Luke, Kent Smith, John Abbott, Peter Lorre Jr. (he was actually an imposter named Eugene Weingand who had once been taken to court by Lorre for fraudulent use of his name; after Lorre's death, Weingand was back to his old tricks, using his fake name to get acting roles).

May 22—KING KONG VS. GODZILLA (1963) Rerun.

May 29—PANIC ON THE 5:22 (1974) Gang of thugs hijacks a commuter train for the purpose of robbing its wealthy passengers. This made-for-tv movie is a routine crime drama, and it is unknown why it aired on *The World Beyond*. It can only be assumed that a KPHO programmer, in a rushed moment, mistook it for a horror movie. With Laurence Luckinbill, Lynda Day George, Dennis Patrick, Andrew Duggan, Ina Balin, Bernie Casey.

June 5—RETURN OF THE GIANT MONSTERS (1966) Gamera, the giant flying turtle, battles Gaos, a Pterodactyl who shoots laser beams out of his mouth. Like some other Japanese monster movies, the American distributor later changed, and retitled this GAMERA VS. GAOS.

June 12—COUNT YORGA, VAMPIRE (1970) A sophisticated vampire puts the bite on people in then-contemporary Los

Angeles. This, and several other films, represented a failed attempt by American International Pictures to turn Robert Quarry into a new horror star. Also with Roger Perry, Michael Murphy. The direct sequel, RETURN OF COUNT YORGA, was never shown on *The World Beyond*.

June 19—THE MAD GENIUS (1931) A sinister Svengali-like dance teacher completely controls the life of his adopted son. With John Barrymore, Marian Marsh, Donald Cook, Frankie Darro, Luis Alberni. Directed by Michael Curtiz. Quite good and fascinating to watch; not really a horror film but came to be regarded as one by some in later years because of its title and the realization that Boris Karloff has an unbilled bit part in it near the beginning. Probably not realizing the circumstance, Stu Tracy named Karloff as the "star" in his World Beyond introduction.

June 26—WAR OF THE GARGANTUAS (1967) Rerun.

July 3—GENESIS II (1973) Rerun.

July 10—KONG ISLAND (1978) Mad scientist turns apes into robots. Rock bottom in all departments. With Brad Harris, Marc Lawrence. Also released under the title KING OF KONG ISLAND.

July 17—DR. WHO AND THE DALEKS (1965) Film version of a long-running BBC tv series which has existed in different forms over many years. Dr. Who and his grandchildren find themselves in a futuristic city where they battle Daleks (malevolent robots). With Peter Cushing, Roy Castle.

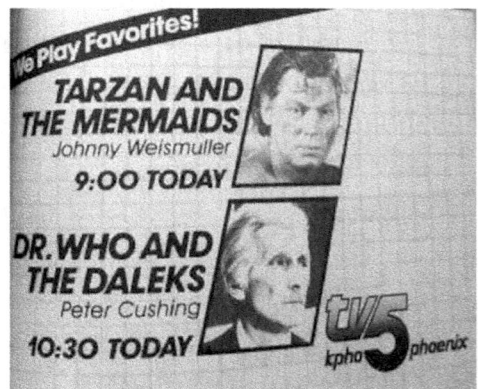

We Play Favorites!

TARZAN AND THE MERMAIDS
Johnny Weismuller
9:00 TODAY

DR. WHO AND THE DALEKS
Peter Cushing
10:30 TODAY

kpho **5** phoenix

There came a time when KPHO stopped using the names of their movie programs in their advertising, although they were still retained on the air.

July 24—GARGOYLES (1972) Rerun.

July 31—BLOOD FROM THE MUMMY'S TOMB (1971) The curse of an Egyptian princess follows the defilers of her tomb. A Hammer Films production. With Andrew Keir, Valerie Leon, George Coulouris.

August 7—THE BAT PEOPLE (1974) A man begins to turn into a bat after being bitten by one. With Stewart Moss, Marianne McAndrew, Michael Pataki, Arthur Space.

August 14—VARAN THE UNBELIEVABLE (1962) Rerun.

August 21—NO SURVIVORS, PLEASE (1963) German film about a reporter who suspects space aliens are infiltrating civilization. With Robert Cunningham, Maria Perschy.

August 28—THE GREEN SLIME (1969) Scaly monsters with tentacles and lots of beady eyes attack a spaceship, and the astronauts still find time to fight over the lone female crew member! A U.S.-Japanese co-production. Pretty bad, but the ludicrous monsters almost make it fun to watch. With Robert Horton, Richard Jaeckel, Luciana Paluzzi.

September 4—SON OF SINBAD (1955) "Arabian Nights" adventure yarn with Sinbad the Sailor captured by an evil caliph. With Dale Robertson, Sally Forrest, Vincent Price, Lili St. Cyr. Possibly scheduled on *The World Beyond* because of Price's presence and the fact that, due to some other movies, Sinbad had become identified with monsters in later years.

September 11—THE CREEPING TERROR (1964) Rerun.

September 18—MADHOUSE (1974) A has-been horror star, attempting a comeback, becomes a suspect when people on the set of his new movie start being murdered. With Vincent Price, Peter Cushing, Robert Quarry, Adrienne Corri. Sounds like it can't miss, but unfortunately it does.

September 25—ISLE OF THE DEAD (1945) One of producer Val Lewton's acclaimed horror films, this one about a Greek Island beset by a terrible plague. Directed by Mark Robson. With Boris Karloff, Ellen Drew, Jason Robards Sr.

October 2—I WALKED WITH A ZOMBIE (1943) Nurse visiting a Carribean Island discovers voodoo and zombies are afoot. Produced by Val Lewton and directed by Jacques Tourneur. With Tom Conway, Frances Dee. Regarded by some as one of the best horror movies of the 1940s.

October 9—LADY FRANKENSTEIN (1973) Dr. Frankenstein's daughter creates a monster to use for her sexual pleasure! Really sleazy Italian film, another one hardly appropriate for the generally juvenile audience of *The World Beyond*. Directed by Mel Welles. With Joseph Cotten, Sarah Bey, Mickey Hargitay.

October 16—THE INCREDIBLE MELTING MAN (1977) Astronaut returns from a disastrous space flight and becomes the title creature. With Myron Healy, Burr DeBenning, Alex Rebar.

October 23—THE BODY SNATCHER (1945) In the 19th century, a doctor hires a grave robber to supply him with cadavers for his experiments and lives to regret it. One of producer Val Lewton's acclaimed horror movies, directed by Robert Wise, who, as we know, went on to much bigger things! With Boris Karloff, Bela Lugosi, Henry Daniell.

October 30—QUEST FOR LOVE (1971) Man is transported to another dimension where Earth looks the same, but he has a different identity. With Tom Bell, Joan Collins, Denholm Elliott, Simon Ward.

November 6—DALEKS-INVASION EARTH 2150 A.D. (1966) Sequel to DR. WHO AND THE DALEKS finds the good doctor traveling to the future again to battle evil robots. With Peter Cushing, Andrew Keir.

November 13—DEATHSPORT (1978) In the bleak, post-Apocalypse future, a nomadic leader battles an evil motorcycle gang. With David Carradine, Claudia Jennings, Richard Lynch, Jesse Vint, William Smithers. Jennings died in a car crash at the age of 29 not long after this was made.

November 20—THE NIGHT STALKER (1972) Reporter Carl Kolchak discovers a vampire causing mayhem in modern Las Vegas. Very popular made-for-tv movie which spawned a sequel (THE NIGHT STRANGLER) and then a short-lived tv series which became a cult favorite after it went off the air. With Darren McGavin, Carol Lynley, Simon Oakland, Barry Atwater, Ralph Meeker, Claude Akins, Kent Smith, Elisha Cook Jr.

November 27—AT THE EARTH'S CORE (1976) A two-man expedition burrows to the hollow center of the Earth and discovers a lost tribe and prehistoric monsters. With Doug McClure, Peter Cushing, Caroline Munro.

December 4—YONGARY, MONSTER FROM THE DEEP (1968) Rerun.

December 11—KONGA (1961) Rerun.

December 18—THE ZOMBIES OF SUGAR HILL (1974) The only "blaxploitation" film of the 1970s to air on *The World Beyond*. Hip black woman strikes a deal with the demon Baron Samedi to unleash a squad of zombies to take revenge upon the white men who killed her fiancée. This was a tv retitle of a film that had been released in theatres simply as SUGAR HILL. It was very much aimed at black audiences. With Marki Bey, Robert Quarry, Don Pedro Colley.

December 25—THE MAGIC SERPENT (1966) Rerun.

1983

January 1—HAUNTS OF THE VERY RICH (1972) Group of people at an Island resort slowly come to realize they're dead and this may be Hell! A made-for-tv movie. With Lloyd Bridges, Edward Asner, Cloris Leachman, Anne Francis, Robert Reed, Donna Mills, Moses Gunn.

January 8—THE DUNWICH HORROR (1970) A warlock kidnaps a college student and plans to sacrifice her to the Devil. A loose adaptation of an H.P. Lovecraft story. With Sandra Dee (trying to change her image), Dean Stockwell, Ed Begley, Sam Jaffe, Lloyd Bochner. KPHO ran this in place of David Cronenberg's 1975 horror film THEY CAME FROM WITHIN which had already been announced and scheduled; it is reasonable to assume someone tipped off the station's programmers that the film's heavy sexual content made it inappropriate for *The World Beyond*'s generally juvenile audience (though one might argue that THE DUNWICH HORROR wasn't much of an improvement!). KPHO later rescheduled THEY CAME FROM WITHIN on the *TV5 Late Late Show*, which generally aired around 2:30a.m.

January 15—GAMERA—SUPER MONSTER (1980) The friendly giant flying turtle of Japanese monster movies is back, though most of his scenes are stock footage from earlier films! Meanwhile, a little boy meets the "Kung Fu Women of Venus" and helps them battle aliens! Easily the worst of the Gamera series, although it is so bad it's kind of fun to watch.

January 22—PHANTOM OF THE RUE MORGUE (1954) Rerun of a film that had not been seen on *The World Beyond* since 1970.

January 29—THE HAUNTED PALACE (1963) Man inherits the old family castle and moves in, and becomes possessed by the spirit of his ancestor who was burned at the stake centuries earlier. Roger Corman directed this disappointing film, which is marred by a script that has the characters never realizing what is happening, so they don't know what they're fighting. With Vincent Price, Debra Paget, Lon Chaney Jr., Elisha Cook Jr., Leo Gordon.

February 5—KILLER GRIZZLY (1976) A tv retitle of a film that was released in theatres as simply GRIZZLY. A marauding bear attacks tourists in a National Park. I suppose it's kind of a monster movie, but not really, since there is nothing supernatural about the beast. With Christopher George, Andrew Prine, Richard Jaeckel.

February 12—SCREAM OF THE WOLF (1974) Small town law enforcement tries to solve a series of murders that may have been committed by a werewolf! Most viewers of this film come away disappointed by the ending which reveals this to be just another murder mystery. Produced and directed by Dan Curtis, with a screenplay by Richard Matheson, both of whom have done much better. With Clint Walker, Peter Graves, Jo Ann Pflug, Philip Carey, Vernon Weddle.

February 19—BEN (1972) Direct sequel to WILLARD (1971) which never aired on *The World Beyond*. Sick little boy befriends the rodent leader of the army of bloodthirsty rats from the first film. With Lee Harcourt Montgomery, Joseph Campanella, Arthur O'Connell, Rosemary Murphy, Meredith Baxter. The young Michael Jackson's hit song, "Ben", had its origin here, believe it or not.

February 26—HORROR AT 37,000 FEET (1973) Rerun.

March 5—BEAST FROM 20,000 FATHOMS (1953) Rerun.

March 12—MONSTER ZERO (1968) Rerun.

March 19—BATTLE OF THE WORLDS (1961) Rerun.

March 26—CRY OF THE BANSHEE (1970) In the 16th century, evil witch Oona summons a demon to take revenge on the witch-hunter who slaughtered her children. Gruesome and largely incoherent film. With Vincent Price, Elisabeth Bergner, Hugh Griffith.

April 2—THE ALIEN FACTOR (1977) Rerun.

April 9—STAR PILOT (1967) Rerun.

April 16—WOMEN OF THE PREHISTORIC PLANET (1965) Rerun.

April 23—BRIDE OF THE GORILLA (1951) Rerun.

April 30—DESTROY ALL MONSTERS (1968) Rerun.

May 7—FOOD OF THE GODS (1976) Rerun.

May 14—THE ABOMINABLE DR. PHIBES (1971) Disfigured madman kills off the team of surgeons he blames for the death of his wife, in creative ways that mirror the Biblical ten plagues upon Egypt. One of the wildest, over-the-top horror films ever made, and very enjoyable on that level. With Vincent Price, Joseph Cotten, Hugh Griffith, Terry-Thomas.

May 21—BAD RONALD (1974) Made-for-tv movie in which a teenager kills a little girl and hides in a secret room in his mother's house. When the old lady dies, he stays there after a new family moves in, spying on them, and slowly going mad. While nothing graphic is really shown due to the tv constraints of the era, this is still one of my top nominees for scuzziest film to be shown on *The World Beyond*. With Scott Jacoby, Pippa Scott, John Larch, Kim Hunter, John Fiedler, Dabney Coleman, Lisa Eilbacher, Ted Eccles, Aneta Corsaut.

May 28—SUPERARGO (1968) A caped, wrestling superhero battles a mad scientist who is turning athletes into robots! An Italian-Spanish co-production, reportedly part of a series! With Guy Madison, Ken Wood.

June 4—THE SPELL (1977) A bullied teenage girl uses supernatural powers to take revenge on her tormentors. A made-for-tv rip-off of 1976's CARRIE. With Lee Grant, James Olson, Susan Myers, Helen Hunt (yes, *the* Helen Hunt, before she hit stardom).

June 11—JENNIFER, THE SNAKE GODDESS (1978) A bullied teenage girl uses supernatural powers to take revenge on her tormentors—yes, the same plot as THE SPELL, except she summons snakes to kill them all off. TV retitle of a film that was called just JENNIFER in theatres. With Lisa Pelikan, Bert Convy, Nina Foch, John Gavin, Jeff Corey.

June 18—DON'T BE AFRAID OF THE DARK (1973) Rerun.

June 25—TIME TRAVELERS (1976) Rerun.

July 2—NIGHT OF THE COBRA WOMAN (1972) Philippine-made exploitation movie about a woman who can turn herself into a cobra, needing snake venom and constant sex to stay young. And you think you have problems? With Joy Bang, Marlene Clark.

July 9—RODAN (1956) Rerun.

July 16—WESTWORLD (1973) Rerun.

July 23—TRILOGY OF TERROR (1975) A made-for-tv horror film with three short stories, all starring Karen Black. Film is remembered exclusively for the final story, in which she is attacked by a bloodthirsty Zuni doll (political correctness would probably prevent this film from being made today). The camera also drools over Karen Black's gorgeous legs and anatomy throughout. Produced and directed by Dan Curtis and written by Richard Matheson. Also with Robert Burton, John Karlen, George Gaynes, Gregory Harrison.

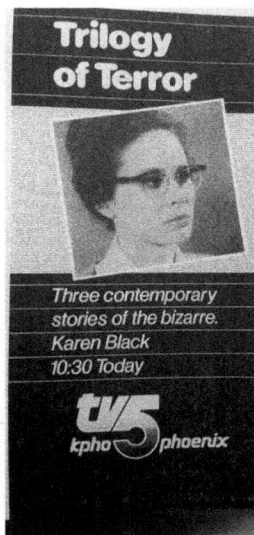

Trilogy of Terror

Three contemporary stories of the bizarre.
Karen Black
10:30 Today

tv5
kpho phoenix

July 30—TERROR OF MECHAGODZILLA (1978) Rerun.

August 6—SNOW CREATURE (1954) An abominable snowman is brought to the city where he escapes and causes mayhem. Directed by W. Lee Wilder. With Paul Langton, Leslie Denison.

August 13—PHANTOM FROM SPACE (1953) An invisible alien wreaks havoc at an observatory. Directed by W. Lee Wilder. With Ted Cooper, Noreen Nash.

August 20—THE MAN WITHOUT A BODY (1957) Scientists revive the talking head of Nostradamus, who still makes accurate predictions while attached to tubes on a table! Nowhere near as much fun as it sounds. Directed by W. Lee Wilder. With Robert Hutton, George Coulouris, Julia Arnall.

August 27—THE OMEGANS (1968) A man takes his revenge on his wife and her lover by forcing them to bathe in a radioactive jungle river. W. Lee Wilder was Billy Wilder's untalented brother, and this was the fourth film by him in a row to be shown on *The World Beyond*! KPHO must have acquired a package of them. With Ingrid Pitt, Keith Larsen.

September 3—THE COLOSSUS OF NEW YORK (1958) Rerun.

September 10—KILLERS FROM SPACE (1954) Space aliens kidnap a nuclear scientists and force him to help with their plans to invade Earth. Yet another film made by the indefatigable W. Lee Wilder. With Peter Graves, James Seay.

September 17—THE OBLONG BOX (1969) In the African jungle, a British aristocrat is targeted for revenge by his deformed brother after being buried alive (!). The title was borrowed from an Edgar Allen Poe story, but this has very little to do with Poe. Despite a good cast, the film is largely incoherent. With Vincent Price, Christopher Lee.

September 24—GODZILLA'S REVENGE (1967) Rerun.

October 1—THE BEES (1978) A large strain of killer bees threatens to decimate the world! Low-budget variation on a much-filmed idea. With John Saxon, Angel Tompkins, John Carradine (playing a character named Uncle Ziggy!).

October 8—THE CREMATORS (1972) A scientist battles an alien orb that sucks up human beings. Or something like that! With Maria De Aragon, Marvin Howard.

October 15—I MARRIED A MONSTER FROM OUTER SPACE (1958) The title pretty much says it all in this opus about a woman who learns that her new husband has been possessed by an evil space alien! This one actually has some admirers! Tom Tryon, Gloria Talbott, Maxie Rosenbloom.

October 22—THE UNCANNY (1977) A series of short stories depicting how evil cats are, and how they are conspiring against mankind. With Ray Milland, Peter Cushing, Donald Pleasence, Susan Penhaligon, Samantha Eggar, John Vernon.

October 29—CURSE OF THE MUMMY'S TOMB (1964) Rerun.

November 5—THE BLOB (1958) Rerun.

November 12—BEDLAM (1946) Inmates rise up against the brutal keeper of an insane asylum. One of producer Val Lewton's acclaimed horror films of the 1940s, directed by Mark Robson. With Boris Karloff, Anna Lee, Ian Wolfe, Jason Robards Sr.

November 19—THE GORGON (1964) Rerun.

November 26—DINOSAURUS! (1960) Rerun.

December 3—DARK STAR (1974) Rerun.

December 10—INVADERS FROM MARS (1953) Rerun.

December 17—EYES BEHIND THE STARS (1972) Rerun.

December 24—PLANET EARTH (1974) Rerun.

December 31—RETURN OF GIANT MAJIN (1966) Rerun.

1984

January 7—FRANKENSTEIN'S BLOODY TERROR (1968) Rerun. This was shown in place of SCREAM OF THE DEMON LOVER (1971), which had originally been scheduled and announced. Why it was cancelled is unknown. Perhaps someone suggested to KPHO that it was inappropriate for *The World Beyond*'s daytime juvenile audience. Whatever the reason, KPHO never rescheduled SCREAM OF THE DEMON LOVER, not even on their late night programming.

January 14—NIGHT CREATURE (1978) Big game hunter living on his own Island has trouble bagging the leopard that is out to get him. Not really a horror film, but always sold as such from the time of its release. With Donald Pleasence, Nancy Kwan, Ross Hagen.

January 21—THE LAST BRIDE OF SALEM (1974) A young couple battles a Satanic cult. Originally premiered on tv as part of the ABC Afternoon Playbreak series, but was later syndicated in movie packages. With Lois Nettleton, Bradford Dillman.

January 28—BEYOND THE BERMUDA TRIANGLE (1975) A businessman starts his own investigation of the infamous section of sea where countless boats and planes have disappeared. A made-for-tv movie. With Fred MacMurray, Donna Mills, Sam Groom, Dana Plato.

February 4—THE TOMB OF LIGEIA (1964) The last of director Roger Corman's loose Poe adaptations, though American International Pictures tried to keep the series going with other people. After he remarries, a man is haunted by the vengeful spirit of his first wife. Screenplay is by Robert Towne, who went on to become a major Hollywood screenwriter. With Vincent Price, Elizabeth Shepherd.

February 11—MYSTERY OF THE WAX MUSEUM (1933) The original classic about a mad sculptor who uses corpses for his wax figure displays. Directed by Michael Curtiz. With Lionel Atwill, Fay Wray, Glenda Farrell, Frank McHugh. This was thought to be a lost film for decades until a print was discovered in 1968, and it gradually got back into circulation again. Remade as HOUSE OF WAX (1953), which had aired on *The World Beyond* in 1969, and imitated many times.

February 18—HOUSE OF USHER (1960) Young man seeking his fiancée's hand in marriage visits the cursed home of her brother and lives to regret it. The first of director Roger Corman's famous Edgar Allen Poe adaptations. With Vincent Price, Mark Damon.

February 25—THE GIANT SPIDER INVASION (1975) A crashed meteor unleashes a squad of giant spiders! A low-budget attempt at recapturing the 1950s, with very fake-looking monsters. With Steve Brodie, Barbara Hale, Leslie Parrish, Alan Hale Jr.

March 3—NIGHT OF THE BLOOD MONSTER (1972) A Spanish-Italian-German co-production about a brutal witch-hunting magistrate back in Medieval times. Directed by Spanish schlockmeister Jesus Franco, though admittedly, this was one of his better efforts. Also released under the less-misleading title THE BLOODY JUDGE. Quite a grisly film; another one that makes one wonder what KPHO programmers were thinking by booking it on *The World Beyond*. With Christopher Lee, Maria Schell, Leo Genn.

March 10—EQUINOX (1971) Very low-budget film—it is almost a home movie—that was released theatrically to some good notices, mainly for its special effects by Jim Danforth. College students search for a missing professor; discover devil-worshippers and a number of large monsters! With Edward Connell, Barbara Hewitt, Fritz Leiber, and supposedly the uncredited voice on a tape recording is Forrest J. Ackerman.

March 17—THE ASTRO-ZOMBIES (1969) Mad scientist creates a squad of zombies from cadavers. This one often appears on lists of the worst movies ever made. This is the only film by schlock filmmaker Ted V. Mikels to be shown on *The World Beyond*. With Wendell Corey, John Carradine.

March 24—THE CREEPING FLESH (1973) Two brothers vie for credit of the discovery of an ancient skeleton, who proceeds to come to life and turn into a monster. Directed by Freddie Francis. With Peter Cushing, Christopher Lee.

March 31—THE IMMORTAL (1969) Man with rare blood disease that keeps him from aging is pursued by bad guys who want to drain his blood. Made-for-tv film was the pilot for a short-lived tv series which followed. With Christopher George, Barry Sullivan, Carol Lynley, Ralph Bellamy, Jessica Walter.

April 7—FRANKENSTEIN CONQUERS THE WORLD (1966) Rerun.

April 14—THE BOY WHO CRIED WEREWOLF (1973) Child discovers his father has become a werewolf, and can't get anyone to believe him. With Kerwin Mathews, Elaine Devry, Robert J. Wilke.

April 21—WHO SLEW AUNTIE ROO? (1972) Two children decide the local crazy old lady is a witch, and murder her at the end of the film. Intended as a twisted variation of Hansel and Gretel. Co-written by Jimmy Sangster and directed by Curtis Harrington. With Shelley Winters, Mark Lester, Ralph Richardson, Hugh Griffith, Lionel Jeffries. Such a film likely would not be made today in this fashion, inasmuch as it has become unpopular in recent years to depict children negatively or villainously.

April 28—WHEN WORLDS COLLIDE (1951) When another planet is about to destroy the Earth by smashing into it, scientists build a rocketship to carry a small number of people to another planet so that the human race may survive. George Pal produced this film, which won an Academy Award for Best Special Effects. With Richard Derr, Barbara Rush, John Hoyt.

May 5—THE LEGEND OF HELL HOUSE (1973) Psychic researchers agree to spend a week at a house known to be inhabited by evil spirits! Written by Richard Matheson. With Pamela Franklin, Roddy McDowall, Gayle Hunnicutt, Clive Revill.

The Legend of Hell House

Starring Pamela Franklin.
10:30 Saturday

tv5
KPHO · PHOENIX

May 12—FROGS (1972) Rerun.

May 19—MURDERS IN THE RUE MORGUE (1971) Another variation of the Poe story about mysterious murders that seem to have been committed by an ape. With Jason Robards, Herbert Lom, Lilli Palmer, Adolfo Celi, Maria Perschy, Christine Kaufman, Michael Dunn.

May 26—GODZILLA VS. THE COSMIC MONSTER (1974) Rerun.

June 2—THE BEAST WITH FIVE FINGERS (1946) A strangler seems to be pursued by his victim's disembodied hand. Horror film fans continue to be disappointed by the revelation at the end that the hand is hallucinatory. Screenplay by Curt Siodmak, and directed by Robert Florey. With Robert Alda, Andrea King, Peter Lorre, J. Carrol Naish.

June 9—DR. PHIBES RISES AGAIN (1972) Sequel to THE ABOMINABLE DR. PHIBES finds the madman seeking an ancient elixir to restore life to his dead wife. Not as good as the first film, but still fun. With Vincent Price, Robert Quarry, Fiona Lewis, Peter Cushing, Hugh Griffith, Terry-Thomas.

June 16—CONQUEST OF SPACE (1955) A saboteur tries to stop the first manned flight to Mars. Is it the captain? Produced by George Pal. With Walter Brooke, Eric Fleming, William Hopper, Phil Foster, Mickey Shaughnessy, Ross Martin, Benson Fong.

June 23—SCREAMERS (1981) On a deserted Island, a mad scientist has been working to create mutant fish people! An Italian film, originally titled ISLAND OF THE FISH MEN. The American distributor went to the expense of filming a new opening sequence with Cameron Mitchell and Mel Ferrer; otherwise, this stars Barbara Bach, Richard Johnson, Joseph Cotten. This was the first film from the 1980s to air on *The World Beyond*.

June 30—MASQUE OF THE RED DEATH (1964) Another of director Roger Corman's Poe adaptations, as a Medieval prince throws a devil-worshipping orgy in his castle while the countryside is being decimated by a plague. With Vincent Price, Hazel Court, Jane Asher, Patrick Magee, Nigel Green.

July 7—GODZILLA ON MONSTER ISLAND (1972) Rerun.

July 14—EMPIRE OF THE ANTS (1977) Rerun.

July 21—THE RAVEN (1963) A battle of wits between two Medieval sorcerers, one good, one evil. Directed by Roger Corman, though despite title, it really has nothing to do with Edgar Allen Poe. Good fun overall, mostly played for laughs. With Vincent Price, Peter Lorre, Boris Karloff, Hazel Court, Jack Nicholson (several years before he became a big star).

THE RAVEN
Vincent Price,
Peter Lorre and
Boris Karloff
10:30 Today

tv5

KPHO · PHOENIX

July 28—SOMETHING EVIL (1972) Family moves into an old farmhouse, and are beset by evil demons! A made-for-tv movie, directed by Steven Spielberg very early in his career—becoming a living legend was still ahead of him. With Sandy Dennis, Darren McGavin, Johnny Whitaker, Ralph Bellamy, Jeff Corey, Bruno VeSota.

August 4—PHASE IV (1974) Rerun.

August 11—ALL THE KIND STRANGERS (1974) Made-for-tv movie about a group of orphans who kidnap motorists, trying to find some who will be willing to stay on as their parents, and kill those who refuse! Another film with a villainous depiction of children—something that isn't done anymore in today's movies. With Stacy Keach, Samantha Eggar, John Savage, Robby Benson.

August 18—RACE WITH THE DEVIL (1975) After accidentally witnessing a human sacrifice by a Satanic cult, two vacationing couples have to take it on the lam with the cultists in hot pursuit.

Most viewers think this film is terrible; I am a dissenter, as I thought it delivers the goods. Judge for yourself. With Peter Fonda, Warren Oates, Loretta Swit, Lara Parker, R.G. Armstrong.

August 25—CRASH! (1977) Old man plots to murder his wife, while she uses psychic powers to make a demonic car drive around killing people! Or something like that. This very incoherent movie is not easy to find today, and it is no relation to *two* later films of the same name that are much better known. With Jose Ferrer, Sue Lyon, John Ericson, Leslie Parrish, John Carradine.

September 1—THE PSYCHIC (1979) Woman has visions of murders that have not occurred yet, in this Italian film directed by Lucio Fulci who became well-known in Italy for directing excessively gory horror films. With Jennifer O'Neill, Gabriele Ferzetti, Gianni Garko.

September 8—GODZILLA VS. MEGALON (1973) Rerun.

September 15—PARTS:THE CLONUS HORROR (1979) The U.S. Government plots to start cloning the population! With Peter Graves, Dick Sargent, Keenan Wynn.

September 22—PLAGUE (1978) A deadly pandemic sweeps across Canada! Not really a horror film per se, but exploitative enough that it has often been regarded as one. With Daniel Pilon and Kate Reid, both of whom would later have recurring guest roles on the famed DALLAS tv series.

September 29—THE TERRORNAUTS (1967) Rerun of a film that had not been broadcast on *The World Beyond* since 1974.

October 6—SCREAM OF THE WOLF (1974) Rerun.

October 13—REPTILICUS (1961) Rerun.

October 20—ASYLUM (1972) Four short supernatural stories, all written by Robert Bloch. Directed by Roy Ward Baker. With Richard Todd, Peter Cushing, Herbert Lom, Barbara Parkins, Sylvia Syms, Robert Powell, Britt Ekland, Patrick Magee, Charlotte Rampling.

October 27—THE MASK OF FU MANCHU (1932) The evil Oriental genius sets out to find the sword of Genghis Khan, so he can summon its powers to take over the world. Today, all of the Fu Manchu stories and movies are considered racist, but if you can overlook that, this entry is fascinating to watch! This was the only Fu Manchu movie to ever air on *The World Beyond*. With Boris Karloff, Lewis Stone, Karen Morley, Myrna Loy, Jean Hersholt.

November 3—KING KONG VS. GODZILLA (1963) Rerun.

November 10—TALES OF TERROR (1962) Three short stories, loosely based on the tales of Edgar Allen Poe. Directed by Roger Corman, and quite good. Written by Richard Matheson. With Vincent Price, Peter Lorre, Basil Rathbone, Debra Paget.

November 17—THE DARK (1979) A space alien roams around Santa Monica, killing people. With William Devane, Cathy Lee Crosby, Richard Jaeckel, Keenan Wynn, Vivian Blaine.

November 24—THE NIGHT STRANGLER (1973) Made-for-tv sequel to THE NIGHT STALKER finds reporter Cark Kolchak tracking a murderer who seems to be immortal. With Darren McGavin, Jo Ann Pflug, Simon Oakland, Richard Anderson, Margaret Hamilton (!), John Carradine, Wally Cox. This was written by Richard Matheson and directed by Dan Curtis. The cult NIGHT STALKER tv series followed.

December 1—SNOWBEAST (1977) Bigfoot terrorizes a ski resort in this made-for-tv potboiler. With Bo Svenson, Yvette Mimieux, Clint Walker, Robert Logan, Sylvia Sidney.

December 8—INVASION FROM INNER EARTH (1972) A small group of survivors hides out in a Canadian farmhouse after aliens take over. Extremely low-budget, and not easy to find today. With Paul Bentzen, Nick Holt.

December 15—CURSE OF THE SWAMP CREATURE (1966) Rerun.

December 22—LOCUSTS (1974) Farmers desperately try to save their crops from a large swarm of locusts heading in their direction. Pretty much a straight drama, apparently mistaken for a horror film by KPHO programmers. With Ben Johnson, Ron Howard, Belinda Belaski.

December 29—FANTASTIC INVASION OF PLANET EARTH (1966) Rerun.

1985

January 5—THE DEVIL'S DAUGHTER (1973) Tv movie about a Satanic cult who kidnaps a young woman for the purpose of marrying her to a demon. With Belinda Montgomery, Shelley Winters, Joseph Cotten, Jonathan Frid, Robert Foxworth, Abe Vigoda.

January 12—DESTROY ALL MONSTERS (1968) Rerun.

January 19—THE SKULL (1965) Effective shocker about a professor who becomes possessed by the skull of the notorious Marquis De Sade. With Peter Cushing, Christopher Lee, Patrick Wymark, Nigel Green, Michael Gough, George Coulouris, Patrick Magee.

January 26—SON OF KONG (1933) Rerun.

February 2—DALEKS:INVASION EARTH 2150 A.D. (1966) Rerun.

February 9—GODZILLA VS. THE SMOG MONSTER (1972) Rerun.

February 16—I WALKED WITH A ZOMBIE (1943) Rerun.

February 23—ISLE OF THE DEAD (1945) Rerun.

March 2—RETURN OF THE GIANT MONSTERS (1966) Rerun.

March 9—THE BAT PEOPLE (1974) Rerun.

March 16—THE INCREDIBLE MELTING MAN (1977) Rerun.

March 23—I, MONSTER (1971) Rerun.

March 30—X-THE MAN WITH THE X-RAY EYES (1963) Rerun.

April 6—THE INCREDIBLE TWO-HEADED TRANSPLANT (1971) Rerun.

April 13—YONGARY, MONSTER FROM THE DEEP (1968) Rerun.

April 20—TERROR IN THE WAX MUSEUM (1973) A mad killer in London seems to be using the local wax museum as his base. Low-grade thriller, but a great cast of veterans almost succeeds in putting it over. With Ray Milland, Broderick Crawford, Elsa Lanchester, John Carradine, Maurice Evans, Louis Hayward, Patric Knowles.

April 27—GOOD AGAINST EVIL (1977) Rerun.

May 4—HOUSE OF USHER (1960) Rerun.

May 11—HUMANOIDS FROM THE DEEP (1980) Sea monsters come ashore, rape women, and commit other acts of gory mayhem. A completely over the top movie that surely had to have been very heavily edited for daytime broadcast tv during this era—including a particularly grotesque climax. With Doug McClure, Ann Turkel, Vic Morrow. Logan Blackwell feels this is, out of a number of contenders in later years, the most inappropriate film to ever be shown on *The World Beyond*.

May 18—HORROR HOUSE (1969) Rerun.

May 25—COMEDY OF TERRORS (1964) Rerun.

June 1—CREATURE WITH THE BLUE HAND (1967) Rerun.

June 8—CAT PEOPLE (1942) Perhaps the best known of producer Val Lewton's famous horror films. A young woman's jealous rages may turn her into a cat. Directed by Jacques Tourneur. With Simone Simon, Kent Smith, Tom Conway, Jack Holt. The sequel, CURSE OF THE CAT PEOPLE, had aired on *The World Beyond* in 1972.

June 15—THE DEVIL'S RAIN (1975) A group of Satan worshippers attacks a family that possesses an ancient book they wish to acquire. This one has often gotten very negative reviews, but I have always found it to be quite effective and atmospheric—albeit not really appropriate fare for *The World Beyond*'s juvenile audience! There is a great climax where the entire cast melts on screen. With Ernest Borgnine, Eddie Albert, Ida Lupino, William Shatner, Keenan Wynn, John Travolta (film debut in a small role before he hit the big leagues), Woodrow Parfrey, Anton Szandor Lavey (the real life head of the Church of Satan, appearing in a bit part—he also acted as the film's technical advisor).

June 22—UNKNOWN WORLD (1951) Rerun.

June 29—UP FROM THE DEPTHS (1979) A huge prehistoric fish snacks on swimmers at a tropical resort. One of many rip-offs of JAWS from the era, with the slight novelty that the shark is, in fact, some kind of monster this time around. Directed by Charles B. Griffith. With Sam Bottoms, Susanne Reed, Virgil Frye.

The Blob

Steve McQueen has a mess on his hands, today at 10:30.

tv5 kpho phoenix

July 6—DINOSAURUS! (1960) Rerun.

July 13—KILLER GRIZZLY (1976) Rerun.

July 20—THE BLOB (1958) Rerun.

July 27—GODZILLA ON MONSTER ISLAND (1972) Rerun.

August 3—BILLY THE KID VS. DRACULA (1966) Rerun of a film that had not been shown on *The World Beyond* since 1969!

August 10—GAMERA-SUPER MONSTER (1980) Rerun.

August 17—THE MAGIC SERPENT (1966) Rerun.

August 24—YOG, MONSTER FROM SPACE (1971) Rerun.

August 31—GAMERA VS. MONSTER X (1970) Rerun.

September 7—ATTACK OF THE MONSTERS (1968) Rerun.

September 14—CURSE OF THE MUMMY'S TOMB (1964) Rerun.

September 21—BEN (1972) Rerun.

September 28—THE GORGON (1964) Rerun.

October 5—GODZILLA VS. THE COSMIC MONSTER (1974) Rerun.

October 12—LEGEND OF BOGGY CREEK (1976) Rerun.

October 19—FOOD OF THE GODS (1976) Rerun.

October 26—FROGS (1972) Rerun.

November 2—THE LAND THAT TIME FORGOT (1975) Rerun.

November 9—SON OF BLOB (1972) Rerun.

November 16—GODZILLA VS. MEGALON (1973) Rerun.

November 23—AT THE EARTH'S CORE (1976) Rerun.

November 30—FIRST SPACESHIP ON VENUS (1962) Rerun.

December 7—THE ABOMINABLE DR. PHIBES (1971) Rerun.

December 14—DR. PHIBES RISES AGAIN (1972) Rerun.

December 21—OLD DRACULA (1975) Unsuccessful comedy about an elderly vampire who resurrects his dead wife with a blood transfusion, but inadvertently turns her black! Directed by Clive Donner. With David Niven, Teresa Graves. The title is a rip-off of Mel Brooks' popular YOUNG FRANKENSTEIN the year before.

December 28—THE TERRORNAUTS (1966) Rerun.

1986

January 4—ATTACK OF THE 50 FOOT WOMAN (1958) After an encounter with space aliens, woman grows to enormous proportions and seeks revenge on the jerk she's married too! Incredibly, a film that became famous for how bad it is, with awful special effects. With Allison Hayes, William Hudson, Yvette Vickers.

January 11—PLANET OF BLOOD (1966) Rerun.

January 18—INVASION (1964) Rerun.

January 25—THE TERROR (1963) Rerun.

February 1—THE THING (1951) Rerun.

February 8—JOURNEY TO THE SEVENTH PLANET (1961) Rerun.

February 15—HOUSE ON HAUNTED HILL (1959) Rerun.

February 22—LITTLE SHOP OF HORRORS (1960) Director Roger Corman's original, near-legendary low-budget opus that was turned into a Broadway musical many years later! Skid-row inhabitant Seymour Krelboined feeds people to a carnivorous talking plant (his name was changed to Krelborn for the later play). Written by Charles B. Griffith, who had to sue to get credit for the story when the play premiered. With Jonathan Haze, Jackie Joseph, Mel Welles, Dick Miller, Jack Nicholson (small role at the beginning of his career).

March 1—THE GIANT BEHEMOTH (1959) Rerun.

March 8—THE ANGRY RED PLANET (1959) Rerun.

March 15—CASTLE OF TERROR (1964) Rerun.

March 22—THE CYCLOPS (1957) Search party, trying to find a woman's missing husband in Mexico, discovers he has been turned into a giant one-eyed monster by radiation! Produced and directed by Bert I. Gordon. With James Craig, Gloria Talbott, Lon Chaney Jr., Tom Drake.

March 29—BARON BLOOD (1972) Italian film with the reincarnation of a sadistic nobleman coming back to continue killing and torturing people. Directed by Mario Bava. With Joseph Cotten, Elke Sommer, Sarah Bay, Massimo Girotti.

April 5—BELA LUGOSI MEETS A BROOKLYN GORILLA (1952) Duke Mitchell and Sammy Petrillo were a low-grade, untalented comedy team who went out of their way to imitate Dean Martin and Jerry Lewis (the real Martin and Lewis reportedly tried to take legal action against them at one point). In this movie, regarded as one of the worst of the worst, bumblers Mitchell and Petrillo visit the jungle and find mad scientist Bela Lugosi hard at work. Even though he had sadly starred in many bad movies, this was a new career low for Lugosi.

April 12—CURSE OF THE SWAMP CREATURE (1966) Rerun.

April 19—THE INDESTRUCTIBLE MAN (1956) Rerun.

April 26—ATRAGON (1963) Japanese sci-fi, produced by Toho Studios, about the crew of a vessel that can act as both an airplane and a submarine going into action when the world is threatened by the undersea kingdom of Mu. Fun stuff.

May 3—THE PEOPLE THAT TIME FORGOT (1977) Rerun.

May 10—WORLD WITHOUT END (1956) Rerun.

May 17—THE BEAST MUST DIE (1974) Rerun.

May 24—EMPIRE OF THE ANTS (1977) Rerun.

May 31—HANDS OF A STRANGER (1962) Pianist whose hands are destroyed in an accident has the hands of an executed murderer grafted onto him. Soon, he seeks revenge on the people responsible for this incident. Unofficial variation of the old HANDS OF ORLAC story, which has been filmed many times. With James Stapleton, Irish McCalla, Paul Lukather, Sally Kellerman (a number of years before she achieved Hollywood stardom).

June 7—DESTROY ALL PLANETS (1968) In this Japanese movie, space aliens land on Earth and turn into a giant flying squid named Viras! Gamera the turtle comes to the rescue. This should not be confused with the similarly titled DESTROY ALL MONSTERS, which aired on *The World Beyond* multiple times. A change in American distributors in later years resulted in a title change to GAMERA VS. VIRAS.

June 14—THE HYPNOTIC EYE (1960) Rerun of a film that had not been seen on *The World Beyond* since 1968, certainly making it new to most viewers.

June 21—DAUGHTER OF DR. JEKYLL (1957) A series of mysterious murders occurs, and Dr. Jekyll's daughter becomes the prime suspect because she happens to be in the area. Before the film ends, a werewolf enters the story as well! With John Agar, Gloria Talbott, Arthur Shields.

June 28—CREATURE FROM THE HAUNTED SEA (1961) Gangsters, trying to cover up their crimes, are inconveniently pestered by a sea monster. Mostly played for intentional laughs; I found it pretty poor, though many of Roger Corman's fans are fond of it. With Antony Carbone, Betsy Jones-Moreland, Edward Wain (he became famed screenwriter Robert Towne years later!).

July 5—MOON OF THE WOLF (1972) Rerun.

July 12—QUEEN OF OUTER SPACE (1958) Rerun.

July 19—PHASE IV (1974) Rerun.

July 26—THE COLOSSUS OF NEW YORK (1958) Rerun.

August 2—GARGOYLES (1972) Rerun.

August 9—THE NAVY VS. THE NIGHT MONSTERS (1966) Rerun.

August 16—THE BODY SNATCHER (1945) Rerun.

August 23—SNOW CREATURE (1954) Rerun.

August 30—THE UNCANNY (1977) Rerun.

September 6—KILLERS FROM SPACE (1954) Rerun.

September 13—THE LODGER (1944) Tenants suspect the new boarder at their London rooming house may be Jack the Ripper! With Merle Oberon, George Sanders, Laird Cregar, Sir Cedric Hardwicke, Skelton Knaggs.

September 20—CRYPT OF THE LIVING DEAD (1973) While exploring on Vampire Island (is there such a place?), an explorer accidentally resurrects a 13th century female vampire. A Spanish import, directed by Ray Danton, that was also released under the title HANNAH, QUEEN OF THE VAMPIRES.

September 27—THE MAN WITHOUT A BODY (1957) Rerun.

October 4—THE ALIEN FACTOR (1977) Rerun.

October 11—GAMERA VS. MONSTER X (1970) Rerun.

October 18—THE CRIMSON CULT (1968) Rerun.

October 25—ATTACK OF THE MONSTERS (1968) Rerun.

November 1—THE HAUNTED PALACE (1964) Rerun.

November 8—GODZILLA VS. THE SMOG MONSTER (1972) Rerun.

November 15—X-THE MAN WITH THE X-RAY EYES (1963) Rerun.

November 22—INVADERS FROM MARS (1953) Rerun.

November 29—CREATURE WITH THE BLUE HAND (1967) Rerun.

December 6—REPTILICUS (1961) Rerun.

December 13—MURDERS IN THE RUE MORGUE (1971) Rerun.

December 20—CURSE OF THE SWAMP CREATURE (1966) Rerun.

December 27—STAR PILOT (1967) Rerun.

1987

January 3—ATTACK OF THE MONSTERS (1968) Rerun, unusually soon after its previous October 25, 1986 showing.

January 10—I MARRIED A MONSTER FROM OUTER SPACE (1958) Rerun.

January 17—THE BEAST WITH FIVE FINGERS (1946) Rerun.

January 24—CONQUEST OF SPACE (1955) Rerun.

January 31—THE MAN WHO COULD CHEAT DEATH (1959) Man discovers he can stay young forever with regular gland transplants from living donors. A Hammer Studios production, with a screenplay by Jimmy Sangster and directed by Terence Fisher. With Anton Diffring, Hazel Court, Christopher Lee.

February 7—ANDROID (1982) On a space station, a human-like robot rebels against the mad scientist who created him. With Klaus Kinski, Don Opper. Noteworthy as the most recent film to be shown on *The World Beyond*—no other film ever aired that was made later than 1982.

February 14—GODZILLA, KING OF THE MONSTERS (1956) Rerun.

February 21—THE GIANT SPIDER INVASION (1975) Rerun.

February 28—THE RAVEN (1963) Rerun.

March 7—STAR PILOT (1967) Rerun, unusually soon after its previous December 27, 1986 airing.

March 14—THE OMEGANS (1968) Rerun.

March 21—PLAGUE (1978) Rerun.

March 28—PHANTOM FROM SPACE (1953) Rerun.

April 4—THE LOVE WAR (1970) Rerun.

April 11—I, MONSTER (1971) Rerun.

April 18—THE CREMATORS (1972) Rerun.

April 25—MASQUE OF THE RED DEATH (1964) Rerun.

May 2—DR. WHO AND THE DALEKS (1965) Rerun.

May 9—GODZILLA'S REVENGE (1967) Rerun.

May 16—THE PIT AND THE PENDULUM (1961) Rerun.

May 23—THE FOG (1980) A strange mist envelopes a small coastal town, bringing with it vengeful, murderous sea ghosts! Generally good chiller, directed by John Carpenter. With Jamie Lee Curtis, Adrienne Barbeau, Hal Holbrook, Janet Leigh, John Houseman (in a great cameo). There was a big-budget remake in 2005.

May 30—RETURN OF GIANT MAJIN (1966) Rerun.

June 6—THE MASK OF FU MANCHU (1932) Rerun.

June 13—THE DEAD DON'T DIE (1975) Man sets out to clear the name of his brother, who was executed for murder, and this somehow leads him to uncover a plot to conquer the world with zombies! With George Hamilton, Linda Cristal, Ray Milland, Ralph Meeker, Joan Blondell, Reggie Nalder, Yvette Vickers. A made-for-tv movie, written by Robert Bloch and directed by Curtis Harrington.

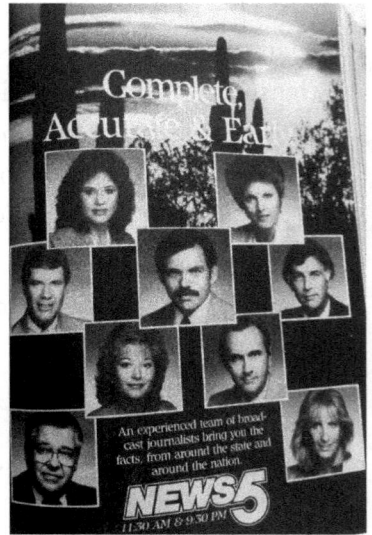

A mid-1980s ad for the KPHO news crew, showing the team including Stu Tracy, who is in the second row on the far left.

June 20—MADHOUSE (1974) Rerun.

June 27—BLOOD FROM THE MUMMY'S TOMB (1972) Rerun.

July 4—THE CONQUEROR WORM (1968) Rerun.

July 11—CASTLE OF THE LIVING DEAD (1963) Rerun of a film that had not been aired on *The World Beyond* since 1970.

July 18—EMPIRE OF THE ANTS (1977) Rerun.

July 25—MYSTERY OF THE WAX MUSEUM (1933) Rerun.

August 1—WAR-GODS OF THE DEEP (1965) Rerun.

August 8—FANTASTIC INVASION OF PLANET EARTH (1966) Rerun.

August 15—LADY FRANKENSTEIN (1973) Rerun.

August 22—EYES BEHIND THE STARS (1972) Rerun.

August 29—FROGS (1972) Rerun.

September 5—SUPERARGO (1968) Rerun.

September 12—THE STRANGLER (1964) Boston authorities try to track down a mad killer who strangles women. With Victor Buono, David McLean, Ellen Corby.

September 19—CRASH! (1977) Rerun.

September 26—FRIGHT (1971) A babysitter is terrorized by a mad killer. With Susan George, Ian Bannen, Honor Blackman, John Gregson.

October 3—THE TOMB OF LIGEIA (1964) Rerun.

October 10—MACABRE (1958) Rerun.

October 17—NIGHT SLAVES (1970) Rerun.

October 24—DON'T BE AFRAID OF THE DARK (1973) Rerun.

October 31—CURSE OF THE FLY (1966) Rerun.

November 7—CONQUEST OF THE PLANET OF THE APES (1972) Rerun.

November 14—DRACULA VS. FRANKENSTEIN (1971) Rerun.

November 21—HUMANOIDS FROM THE DEEP (1980) Rerun of one of the most inappropriate movies to ever be shown on *The World Beyond*.

November 28—DEATH AT LOVE HOUSE (1976) A writer, researching the life of a famous 1930s actress, moves into her house and riles her evil spirit. A made-for-tv movie. With Robert Wager, Kate Jackson, John Carradine, Sylvia Sidney, Joan Blondell, Dorothy Lamour.

December 5—KILLER BEES (1974) Bees kept at a vineyard start buzzing around killing people in this made-for-tv movie, directed by Curtis Harrington. With Gloria Swanson (!), Edward Albert, Kate Jackson, Roger Davis, Craig Stevens.

December 12—WHEN WORLDS COLLIDE (1951) Rerun.

December 19—BATTLE OF THE WORLDS (1961) Rerun.

December 26—MARK OF THE VAMPIRE (1935) A vampire hunter investigates the undead in Eastern Europe—but in one of the most reviled plot twists in film history, it turns out this is simply all a hoax to trick a murderer into confessing to his crime! Until that cop-out occurs, the film is eerie and quite atmospheric. Directed by Tod Browning. With Lionel Barrymore, Elizabeth Allan, Bela Lugosi, Lionel Atwill, Jean Hersholt, Carroll Borland, Donald Meek. This is no relation to the 1957 film of the same name that had aired several times on *The World Beyond*.

1988

January 2—DALEKS-INVASION EARTH 2150 A.D. (1966) Rerun.

January 9—THE INCREDIBLE MELTING MAN (1977) Rerun.

January 16—THE HOUSE THAT WOULD NOT DIE (1970) A woman and her niece inherit an old house and move in, which— as usual in the movies—is a big mistake, as evil spirits live there as well. A made-for-tv movie written by Henry Farrell. With Barbara Stanwyck, Richard Egan, Kitty Winn. This was the last *World Beyond* "premiere"; a film that had not previously aired on the program. The remaining episodes would be all reruns.

January 23—THE CYCLOPS (1957) Rerun.

January 30—WHERE TIME BEGAN (1978) Rerun.

February 6—HOUSE ON HAUNTED HILL (1959) Rerun.

February 13—GODZILLA VS. MEGALON (1973) Rerun.

February 20—LITTLE SHOP OF HORRORS (1960) Rerun.

February 27—YONGARY, MONSTER FROM THE DEEP (1967) Rerun.

March 5—THE HYPNOTIC EYE (1960) Rerun.

March 12—SCREAMERS (1981) Rerun.

March 19—THE ANGRY RED PLANET (1959) Rerun.

March 26—FROM HELL IT CAME (1957) Rerun.

April 2—THE PSYCHIC (1979) Rerun.

April 9—THE GIANT BEHEMOTH (1959) Rerun.

April 16—CASTLE OF TERROR (1964) Rerun.

April 23—FIRST SPACESHIP ON VENUS (1962) Rerun.

April 30—DEATHSPORT (1978) Rerun.

May 7—JOURNEY TO THE SEVENTH PLANET (1961) Rerun.

May 14—RODAN (1956) Rerun.

May 21—KONGA (1961) Rerun.

May 28—GAMERA-SUPER MONSTER (1980) Rerun.

June 4—THE DARK (1979) Rerun.

June 11—*The World Beyond* was apparently pre-empted by the Portland Rose Festival Parade.

June 18—GODZILLA VS. THE COSMIC MONSTER (1974) Rerun.

June 25—DESTROY ALL MONSTERS (1968) Rerun.

That was it. *The World Beyond* vanished from KPHO's programming line-up. There was no announcement or explanation. Although I was a young man in my 20s in 1988, I still occasionally tuned in, because I had a soft spot for such movies, especially those of the Japanese fighting monsters. I recall watching DESTROY ALL MONSTERS that day, even though I had seen it several times over the years….and I still have a memory of thinking it highly unusual that when it ended, Stu Tracy's traditional announcement of what would be shown next week did not appear. I hoped it was just a fluke; it was not.

Somehow, DESTROY ALL MONSTERS seems like a fitting way to end *The World Beyond*. It is unlikely KPHO programmers were even thinking of such a thing; it is doubtful anyone was thinking "What would be a good way to end this?" But since Godzilla and his rubber-suited pals had been such a major part of *The World Beyond* over its lengthy run, it seems appropriate that the final movie would star almost all of them!

So why did KPHO TV5 cancel *The World Beyond* after 24 years? There was never any "official" reason, but looking back on the era, there were several factors in motion. First, the 1980s saw major revolution in the television industry. The VHS videotape exploded onto the market, making it easy and cheap for film lovers to actually own the movies they loved or even just wanted to see. People no longer had to scan issues of TV Guide to see if anything good was coming on; they could access movies far easier. This has only increased in the ensuing years, with DVDs, Blu-Rays, Netflix, and streaming movies on your iPads and cell phones!

Second, the 1980s saw the explosion of cable television, which has also increased in the ensuing years. King Cable had begun slowly in the late 1970s, but by the mid-1980s, the first "super-stations" were appearing, and they had money behind them. They started buying up exclusive rights to virtually everything that was popular with viewers—including old movies and reruns of popular tv shows. Syndicated packages of such material, which had long been the bread and butter of independently owned tv stations nationwide, were rapidly disappearing. Stations that utilized the free airwaves simply could not compete. Gradually, all broadcast stations, including the major networks, got out of the movie business. KPHO slowly started ending its other movie programs as well in the next few years, before being acquired as a CBS affiliate. In addition to *The World Beyond*, I shed a few tears when *Studio 5* ended as well. KPHO's fabled and legendary kid show hosts, Wallace and Ladmo, went off the air at the end of 1989 after a 34 year run, at which time they were the last local personalities of their kind in the nation.

Finally, tastes were changing too. In the 1980s, children and adolescents were being entertained by video games and high-energy film extravaganzas such as RAIDERS OF THE LOST ARK and TOP GUN. Kicking back and watching hoary trash like THE ANGRY

RED PLANET just didn't do it for them anymore. *The World Beyond*'s core audience was gone.

However, for lifelong Arizona residents who grew up on *The World Beyond*, nostalgia for the program has grown in recent years. This nostalgia has culminated in this book you are now holding, and which we thoroughly hope you have enjoyed. Logan Blackwell has a World Beyond appreciation page on Facebook, which is frequently visited by those who loved the program. You can find it at https://www.facebook.com/The-World-Beyond-Kpho-Tribute-171986133007005/?hc_ref=PAGES_TIMELINE&fref=nf

Nostalgia for *The World Beyond* on KPHO-TV5 had led some to fantasize how wonderful it would be for a revival of the show to start up again. While it is a beautiful thought, I am convinced it could never succeed. Today's incredible access to movies, inconceivable in the days of *The World Beyond*, would doom it from the start. The magic of the time is long gone. Likewise, it would only have an audience of those nostalgic for the old show, and it still wouldn't be the same without Stu Tracy's great introductions!

The World Beyond was the product of its era—a wonderful era that today's young people cannot even comprehend. Can you even conceive of your grandchildren squealing with delight over something like THE CREEPING TERROR? Of course not. But for those of us who grew up with *The World Beyond* and KPHO's other movie shows, no one can take away our memories!

AFTERWORD

As one examines the list of movies that aired on *The World Beyond*, it is amazing to note the variety. However, it is just as amazing to note what did not air. Movies were available in syndicated packages to tv stations, but KPHO did not purchase everything that was out there.

For instance, *The World Beyond* carried a lot of Universal Studio's classic monster movies—-but not all of them. Conspicuously absent from this list are SON OF FRANKENSTEIN and HOUSE OF DRACULA, even though KPHO had aired them on *Studio 5* a few times. THE MUMMY'S HAND was shown more than once, but this film's three direct sequels (THE MUMMY'S TOMB; THE MUMMY'S GHOST; and THE MUMMY'S CURSE) never appeared.

A small number of movies produced by the legendary Hammer Studios aired on *The World Beyond*, but not even one of their famed Dracula or Frankenstein films was ever shown. The poverty row studio Monogram produced quite a few very poor horror films with Bela Lugosi—these have long been in the public domain, but only VOODOO MAN and the two outings Lugosi made with the East Side Kids were shown on *The World Beyond*. The films of the infamous Edward D. Wood Jr. never appeared either.

Interestingly, all of Toho Studio's first round of Godzilla movies aired on *The World Beyond* at one time or another (Toho has made Godzilla movies down to the present day, but the first round is generally regarded as the entries filmed between 1954 and 1978). Many other Toho monster flicks aired on *The World Beyond* as well, with only the original MOTHRA (1962) and KING KONG ESCAPES (1968) conspicuously absent.

In later years, *The World Beyond* aired a number of Toei Studio's Gamera films—but not all of them. Most noticeably missing is the very first one, GAMMERA THE INVINCIBLE (1966). GAMERA VS. ZIGRA (1971) also never appeared.

Fans can point to many other films with major horror "stars" that also never aired on the show. While one may regard this as unusual, we must remember that, at the time, there was no real strategy to booking movies on tv. For *The World Beyond*, KPHO's programmers simply were looking for any horror/sci-fi films. They never dreamed people would even remember *The World Beyond* so many years later. But remember we do, and they are happy memories!

APPENDIX

Parker Anderson Interview with Stu Tracy

FOLLOWING IS A TRANSCRIPT of an interview I conducted with former KPHO announcer Stu Tracy on May 12, 2016 at the Deer Valley Airport in Phoenix:

Q: This is Parker Anderson, May 12th, 2016, interview with Stu Tracy at the Deer Valley Airport. Thank you for agreeing to this interview.

A: You're very welcome.

Q: I guess first can you tell me a little about your background, are you a native Arizonan?

A: No, actually, I grew up in Oregon –

Q: Ah.

A: – and when I was 25, I moved to Arizona, had a job opportunity down here that would give me a whole hundred dollars a month more, but that was a great percentage more because I wasn't making anything.

Q: That's great.

A.   It was a move, I was working at a TV station up there in Oregon, in my hometown of Medford. A gentleman I worked with there left and came down to Phoenix, became program manager of a rather obscure radio station down here called KBUZ, and so I moved, he wanted me, he thought I had announcing talent, and he brought me down here to be an announcer, kind of a disc jockey, on this radio station that played music to ride elevators by [laughs]. It was a very obscure station, and after about a year I started looking around right away because I knew I didn't want to stay there –

Q: Uh-huh.

A: – and I had an opportunity to get on with KPHO Radio. At that time, KPHO Radio studios were in Chris-Town, upstairs in the Chris-Town shopping center, and I was hired there in December, I think it was December 1st, 1969.

Q: Wow.

A: So that was my beginning with KPHO. After about a year, I moved over to the TV side; they decided they wanted me on that side. So I was able to make that move and I was there for the next thirty years; in total I was there about thirty-one-and-a-half years.

Q: Right.

A: So I retired in May of 2001.

Q: So when you started at KPHO, it was originally just a straight announcing, did it start limited or were you the Voice of KPHO right away?

A: I was pretty much the Voice, I did what we called the Book every night, put it on the reel tape with stop cues in between each announcement, and that would run throughout the day in sequence and so during all the breaks my voice would be there.

Q: And top of the hour, "KPHO TV-5 Phoenix" –

A: Yeah, every half hour, had to have the official call letters, official station identification, back then. Of course, that didn't fill my day, it didn't take eight hours to do that, and I would do other assorted announcements that were just put on cartridges— carts, we called them—and so I would do that. Like I said, that didn't fill out my day, so they had to find something else for me to do. So I think initially they had me doing sports, and that went on for a year or two—and I have about as much interest in sports as that glass of water does.

Q: You know, I identify. I've never been a sports fan myself.

A: [Laughter] I haven't either, and I guess that surprises some people, but I just, you know, my thing is airplanes, right? And I hang out here at the airport a lot.

Q: Did you ever fly in your younger years? Were you a pilot?

A: Well, my son and I started flying an ultralight in 1980 –

Q: Oh, okay.

A: – and he was fourteen at the time, and he was the best pilot of the group that we were flying with. He learned to fly very quickly, his first flight was a solo because it was a single-seat wingshift ultralight, and he went off to college after a couple of years, and I sold the ultralight because it takes two to put it together, and we'd fold it up and keep it in our garage. But he went off to college, and eventually I started flying with a guy who worked at KNIX Radio. He did the traffic watch on KNIX, Dick Layton (?), then he taught me to fly and I taught him to do the weather. Then my son was in college and so I didn't have anybody to fly with around in the ultralight so I eventually, through Dick Layton, learned to fly and got my pilot's license and started flying Cessnas. So I like to say that I graduated to Cessnas and my son eventually graduated to F-18s; he became a Marine Corps aviator. So that's kind of my extent of flying, I'm still flying to this day, we bought into, as partners, bought into a plane in 1988; we still have it although we bought out our partner.

Q: Wonderful. Now back in those days KPHO was an independent station, I know they ran a lot of movies, it was the movie channel around; and in those days the independent stations—

you know, movie stations today and cable don't do it so much anymore, but they had the intros for movies and you got to do all of those for KPHO.

A: Right, we had movies—I forget what showcases we owned, that we had, including *The World Beyond* and *Movie Time* and there were two or three others.

Q: Oh, there were …

A: You may remember them better than I do.

Q: I probably do, but there were especially a lot of films on the weekends at KPHO.

A: Oh, yeah, we had to fill the time—because we didn't have newscasts on the weekend.

Q: Right. So did they have you record the intros, like, a week at a time, or how often did you have to record all of the week's announcements?

A: Yeah, I had a stack of carts, one being the intro, one an extra-Oh (?) we called it for the breaks, and an open and a close, so I had four carts for every movie. For the weekend ones, I'd do those all at once. Each day's movie, I guess I did them every day.

Q: Okay.

A: Yeah, but we had a stack of carts, two stacks about two feet high, they were about a half-inch thick.

Q: And before we started with the recorder on, you were telling me with the bumpers, "We will return" and stuff, you would change the stars on each one …?

A: Yes, there was usually the name of the movie on a little script they gave me, there was the name of the movie and the stars, the principal stars, usually three or four of them, and I would try to alternate those for each break so the break didn't sound the same: "We now return to …" such-and-such, and "We'll be back …" something, whatever movie, "… right after this!"

Q: Yeah, I remember those. As a child growing up I was quite a film buff, so yeah I'd watch a lot of those, so I heard your voice all the time.

A: Well, you were probably a little kid when I was doing this –

Q: Yes, uh-huh –

A: You're obviously quite a bit younger than I am –

Q: Yeah, well, I'm fifty-two now, so –

A: Okay, yeah, you're the age of my daughter. [Laughter]

Q: You don't have to answer this, but how old are you now?

A: I'm seventy-three.

Q: Oh, are you? Very good.

A: I was twenty-five when I started, twenty-six when I started at KPHO.

Q: Right.

A: Yeah.

Q: You eventually became the newscast's weatherman.

A: Yes, they decided one day that Art Brock, who had been our weatherman –

Q: Yes …

A: No, they wanted me to do the midday show—Art did the evening show for a long time—and he had been at Channel 3, before that Channel 5, and he was Mister Humidity, a great personality, I thought Art did a great job. But he had gray hair, and I guess that's the downfall of all men in local television –

Q: Yeah.

A: National television, that works, the older the better! But if you're on local TV, the younger the better. So I don't know why that is, but it seems to be.

Q: Um-hm.

A: So Art was there for the evening shows, and one day they said, "You're going to do both shows. You're going to work a split shift, come to work twice a day, that's all you mainly have to worry about, that and the movie intros." So that's where I got into that, and actually initially—I need to admit this—I didn't know a lot technically about weather.

Q: Right. That was going to be my next question, so …

A: But a new news director came in, and he says, "You're doing a good job, nobody knows you're not a meteorologist," and most people weren't back in those days, most people on local television were not. He says, "I want you to get a degree in meteorology, well, a degree or at least a certificate in meteorology." At Mississippi State University they had a program—or had—I'm sure they still have—a program called Broadcast Meteorology, and it's almost a four-year course by correspondence; three-and-a half-year, I guess, I forget how many semesters it was. And it was a grueling course. It taught you meteorology. And, though I'm not a degreed meteorologist, taken as a major, but I did get a certificate in meteorology and it taught me a lot— and it taught me a lot that I didn't need to know, because they would put in things just to stretch out the course, they'd have, oh, Statistical Meteorology –

Q: Yes.

A: – it was the most useless course there was, you know?! Nothing anybody would ever use, but they had to fill the time and that made it difficult, 'cause I didn't like Stats.

Q: Well, you know, Jerry Foster has put a book out on his memoirs, and –

A: Oh, you've seen his book?

Q: I've seen his book. Yeah, he says that when he was weatherman for a while, KPHO was on a half-hour earlier and he says now he used to swipe your data; he'd watch yours and then take yours rather than go to the effort.

A: Yeah, 'cause Jerry was not a meteorologist.

Q: Right.

A: And they made him do the weather at Channel 12. And he'd be out working a crime scene or something or searching for somebody in the morning and he had to be there on the air at noon for their 12:00 newscast.

Q: Right.

A: And I was on at 11:30. We had two newscasts, 11:30am and 9:30pm. So Jerry had that one, and in fact I had lunch with Jerry a couple of months ago—we still see each other and I talk to him on the phone. He's a great guy—and he was an incredible pilot, which you can't do any more, but one day he called me up and says, "Hey, why don't you come to lunch with me?" He says, "When you get off the air, come on down to Channel 12 and I'll be getting off the air about then, by the time you get here. So come on down." So I went down and we went up on the roof and got into the helicopter and flew off to his house in Cave Creek –

Q: All right.

A: – for lunch. And that was quite a memorable experience.

Q: I bet it was. Now when you were the Voice, they gave you the information on what to record—do you have any idea … I'm sure that when they purchased the rights to air syndicated packages of movies KPHO had to decide what would fit best on what movie program. Did they have someone specific to do that, well, this movie goes here, this movie goes on that show …?

A: Let's see, I don't think the program director would … I think the program director would delegate that. We had the Promotion

Department, there was a lady there who put those movies into the slots, but I don't remember exactly how that worked.

Q: Yeah. Were you ever any kind of a film buff yourself, or did you never get into that?

A: I never really got into it. I would watch movies, and I prided myself on—I mean, given that we had to break for commercials and at least keep the audience informed—but what the movie was and who was in it, that was my big thing, and I enjoyed doing that.

Q: Right, that's good. I just asked because Mr. Blackwell, who does the Facebook page I mentioned, wanted me to ask if you had any favorite science-fiction movies or anything.

A: You know, one, I wasn't a big movie buff –

Q: Right.

A: – there are some movies that I've enjoyed thoroughly, but I was never into science fiction.

Q: I do remember once in a while, when I was watching as a child, you apparently would go on vacation for a week now and then, and somebody else filled in for you on the announcements. I wondered whether you remembered, off the top of your head, who that was.

A: No … when I would be gone from the weather, they'd bring in the weekend weather guy –

Q: Right.

A: – Ron Merritt, for a long time. But a lot of it I prerecorded. Yeah, occasionally somebody else would have to do the announcements, but for the most part when I was gone for a week I'd have all those things prerecorded, and then if there was some kind of change in the programming, an emergency, they'd put someone else in there for a while. But there was no regular second announcer.

Q: Right. I did ask, because I do remember occasionally—those childhood memories—there'd be a different announcer, but it would only be for, like, a week or so.

A: Yeah, that's probably so, I can't remember who would've done that. In the early days, the guy who'd been the announcer, if you could call it that, was Steve Jenson (sp?), great guy, he was a newscaster we lost a number of years ago, but he would fill in I think in the early days, but later on, I can't remember. And I was only the announcer as such until about, I wanna say, '92.

Q: Yeah, 'cause it was in that vicinity that KPHO became the CBS affiliate –

A: I think that happened in '94 –

Q: Right.

A: – but there was a promotion director that came along, I can't remember her name, she didn't think the on-camera weatherman should be the announcer, it would look like the station couldn't afford more than one person. And in those days, Meredith Corporation was notorious for pinching pennies.

Q: Right.

A: And so what looked like they were trying to save money, was probably the truth.

Q: Right.

A: But, I'm trying to think –

Q: All right.

A: Yeah, but Meredith Corporation was kind of tight, and our salaries reflected that, too. They were not—you know, I like to say, especially in the later years, if I made twice as much as I did, I'd still be making half as much as people thought I did.

Q: Right [laughter], that's the way it often is. I've heard that in media people just assume that because they see you all the time you must be making a lot of money, and that's not always the case.

A: That's not the case at all, especially with Channel 5. I would lobby for a raise, you know, "Munsey's (?) gotta be making blah-blah-blah, so I need a raise." "Oh, yeah, but he delivers more homes than you do." CBS affiliate and all that. [unintelligible]

Q: Right.

A: But this promotion director just decided we needed an extra announcer. So they went through a raft of guys, I mean, there was some kid came in and did it for, I don't know, a month or two, and somebody else would come along, and they never had a station announcer or staff announcer after that that you would recognize, and –

Q: Yeah.

A: But, oh, in '94 we did become the CBS affiliate.

Q: CBS, yeah, and that of course changed the whole thing 'cause then it was all pretty much most of the local programming fell by the wayside and network programming took over.

A: It changed everything, and as the independent we always felt inferior to the network stations, and it came along and it was, like, "Oh, my God, we're going to be a network affiliate now, and isn't that great?!" Well, be careful what you wish for.

Q: Absolutely.

A: It wasn't nearly as great as we thought it would be –

Q: I'm sure it wasn't.

A: – from a personnel standpoint.

Q: And all those years as an independent, KPHO had, I thought, was some really terrific programming.

A: Yeah, so …

Q: Do you remember offhand—maybe you don't, maybe you never did know—who was KPHO's voice before you came on in '69?

A: Well, Steve Jenson for a while; before that, I didn't know the people before that.

Q: Right, sure.

A: No, I don't know who it was. I'm sure it's in the archives there somewhere –

Q: Probably.

A: – but I don't think they had a regular person they called an actual staff announcer, I think various people did it.

Q: Very good, yeah. Let me see, anything else I need to ask? I was kind of a strange child, I got into, I became a movie buff at an early age, and of course that was way in the days before VCR's and recordings ... but I had an audio recorder and I developed a very strange hobby, and if I watched a movie I'd record the intro.

A: Oh, really?

Q: Yeah, so I recorded quite a few movie intros, including lots of KPHO –

A: No kidding!

Q: – and I still have them, and I sent the tapes I had of the old *World Beyond* intros to Mr. Blackwell, who transferred them to CD's so he could put a couple of them on YouTube –

A: Oh, really?

Q: Yeah.

A: No kidding?

Q: No kidding. So, yeah, there are recordings of some of your intros that have survived –

A: I'll be darned.

Q: – because I had such a strange hobby as a child, so ...

A: Yeah. That's amazing.

Q: So if you do YouTube, you might find a couple of those –

A: Yeah, so how would you search?

Q: Well, I think he uses key words like *World Beyond* and *KPHO*, so you could –

A: All right.

Q: – keyword those.

A: Yeah, I'll try that. [laughter] That's amazing. Somebody sent me a link to a YouTube video; I clicked on it and it was one of our old newscasts –

Q: Wow.

A: – and I don't know how it came about, but I think it was Roger Downey, Linda Turley and me, John Brictson on sports, and it was amazing to see that from probably the early 90s.

Q: I bet it was, yes.

A: And how it survived, nobody knows.

Q: It's amazing how some things survive, yeah. Um, let me see here, that probably covers about everything I needed … You're married, you've had children …

A: Yeah, from a previous marriage, I have a son and daughter.

Q: Okay, very good.

A: And they're in their early fifties, my son just turned fifty –

Q: Right.

A: – and he is working on the F-35 project in Washington, DC, he's a manager over a bunch of people. And my daughter's raised four boys in Baltimore, just outside of Baltimore –

Q: Right.

A: – so both kids and all the grandkids are back East, they kind of grew up without us, and most of them are in college now, so …

Q: Yes. Well, I sure thank you very much for doing this interview.

A: Oh, you're very welcome.

Q: I will keep you posted if something indeed does come up and this gets published –

A: All right!

Q: If it becomes a book, I might ask if you'd be interested in writing the foreword to it.

A: Sure.

Q: We'll see what happens.

A: All right.

Retired KPHO meteorologist and announcer Stu Tracy, with author Parker Anderson, following an interview at Deer Valley Airport on May 12, 2016. (Author's collection)